BLUEPRINT 2000

Blueprint 2000

A Conservative Policy towards Employment and Technology in the Next Century

Edited by
Robert Harvey
for the CARE Group of MPs

Foreword by Jim Lester, MP

MACMILLAN
PRESS

© Robert Harvey 1988
Foreword © Jim Lester 1988

First published 1988

Published by
THE MACMILLAN PRESS LTD
Houndmills, Basingstoke, Hampshire RG21 2XS
and London
Companies and representatives
throughout the world

Typesetting by Footnote Graphics, Warminster

Printed in Great Britain by
Richard Clay Ltd, Bungay, Suffolk

British Library Cataloguing in Publication Data
Blueprint 2000: a Conservative policy
towards employment and technology in the
next century.
1. Great Britain. Employment
I. Harvey, Robert
331.12′5′0941
ISBN 0–333–45822–2 (hardcover)
ISBN 0–333–45823–0 (paperback)

This book is dedicated to the memory of **Lord Stockton**, who always looked forward, never back; and to **Iain Macleod**, who achieved so much and had so much more to give.

Conservative Action to Revive Employment (CARE) was founded in 1985 as a group of 120 Conservative MPs concerned with the persistence of high unemployment in an otherwise booming British economy. The views expressed in this book are those of the authors alone, as individual members of CARE, and in no sense express the views of the group as a whole.

Contents

Notes on the Contributors

Tony Baldry, MP is Conservative MP for Banbury, and is parliamentary private secretary to the Minister of State at the Foreign Office. He is a founder member of the Select Committee on Employment.

Sir Philip Goodhart, MP is Conservative MP for Beckenham and is a member of the Select Committee on Employment. He is a former Under-Secretary of State for Defence and for Northern Ireland.

David Grayson worked for an international marketing company before establishing one of Britain's first enterprise agencies in 1980. He was secretary-general of the Democratic Youth Community of Europe for 1979 to 1983, fought Sheffield Brightside in the 1983 election and was Euro-Candidate for Sheffield in 1984. He is a vice-chairman of the Tory Reform Group.

Robert Harvey is former Conservative MP for Clwyd South West and a former member of the House of Commons Select Committee on Foreign Affairs. He was formerly Vice-Chairman of CARE. He is currently an editorialist and columnist on the *Daily Telegraph*, and was previously an assistant editor of the *Economist*.

The Rt Hon Michael Heseltine, MP is Conservative MP for Henley. A former Minister for Aerospace, he has been Secretary of State for the Environment and Defence.

Jim Lester, MP is Conservative MP for Broxtowe. A former Under-Secretary of State for Employment, he is Chairman of CARE and of the Conservative backbench Employment Committee. He is a member of the Foreign Affairs Select Committee.

The Rt Hon Peter Walker, MP is Secretary of State for Wales and Conservative MP for Worcester. A former Minister of Housing, and Local Government, he has been Secretary of State for the Environment, Trade and Industry, Agriculture, and Energy.

Foreword

The success of the post-1979 Conservative government has been undeniable. After a difficult start, during which we can afford to acknowledge such mistakes as an overvalued pound, an over-rigid adherence to narrow monetarist goals, and the premature abandonment of an incomes policy, a cautious yet flexible and ingenious economic policy has been pursued that has given Britain seven years of uninterrupted growth, averaging around 3 per cent a year, faster than our European competitors and even, in some years, than the United States and Japan. Britain, once the sick man of Europe, is now its most robust economy.

Inflation, meanwhile, has been reduced to below 5 per cent for five years, touching as low as 3 per cent in some years. This has not satisfied the Treasury; but it represents an astonishing and consistent achievement for a country which has recently known inflation rates of around 25 per cent and which, in the fifteen years previous to 1979, had rarely seen inflation drop into single figures. The current account of the balance of payments has been consistently in surplus every year since 1979. While the oil surplus was initially responsible for this success, in 1986 even without the oil contribution Britain was in deficit by only £1.5 billion (although, of course, the deficit on manufactured and other goods remains of concern). This about-turn in Britain's economic fortunes deserves the label 'economic miracle'.

The Chancellor is to be congratulated on his fiscal prudence, although we consider that a Public Sector Surplus in 1988 GDP is to err on the side of caution. He is also to be congratulated on his decision substantially to increase public spending on Britain's run-down infrastructure in the autumn of 1986; on his courage in flouting economic orthodoxy by ignoring the narrow money target of Sterling M3, which was allowed to rise by 15 per cent in 1986 without – as the monetarists predicted – re-igniting the fires of inflation, and which is anyway almost uncontrollable in a credit card society; for placing an intelligent reliance upon interest rates to control the money supply, although these are still too high; and for acknowledging the major contribution of inflationary pay settlements to inflation and unemployment. The government as a whole is to be congratulated on its determination to stand up to excess wage demands in the public sector.

It can fairly be said that Britain is overcoming its immediate problems born of the crazy economics of the 1960s and much of the 1970s. The task now, while adhering to sound economics, is to ensure a fairer society: one that unharnesses the energies of those who can help themselves to create wealth for everyone, while doing much more to help those who cannot help themselves. Future Conservative governments must turn to tomorrow's goals: in particular, the huge, persistent and tragic problem of unemployment on a scale that would have been unthinkable fifteen years ago. This book, born of the deliberations of the CARE group of Conservative MPs, attempts to do that. Britain today is undergoing a technological revolution imposing massive changes upon our society that will last well into the twenty-first century; the current unemployment is one symptom of this. The sharp fall in unemployment from more than 3 million to 2.5 million in two years is greatly to be welcomed. All the same, the existing figure represent an unacceptably high level of human misery and wasted resources.

This technological revolution is enormously exciting. It could create a society of much greater wealth, in which the soul-destroying drudgery of the assembly line, of mechanical work, and of much clerical work are things of the past. This technological revolution could – because the new types of assembly are so adaptable – replace the mass production of single-standard goods with quality goods and an infinite variety of services, providing a more diversified and rewarding life for all our people. The challenge of the twentieth century has been to provide a better standard of living for all. The challenge of the twenty-first will be to provide a better quality of life for all. The technological revolution could, in addition, provide everyone with the greater leisure to enjoy this higher quality of life. The opportunity is immense and historic.

Yet if government does not face up to the challenge of managing the social consequences of technological change, the technological revolution could impose strains and misery upon Britain as great as those of the Industrial Revolution. This book is a first attempt to face up to the challenge from a Conservative perspective.

JIM LESTER

1 The New Politics

An industrial revolution is creeping stealthily across the land. It is creeping into traditional offices, where the gentle tapping of word processors has replaced the clatter of manual typewriters and the hum of electric typewriters. It is creeping into car assembly lines, where the whirr and clatter of machinery is being muffled and the shouts of men to each other across the shop floor are replaced by the muted tones of supervisors and technicians, inspecting and monitoring arcades of complicated equipment. It is revolutionising the smaller industries, where business computers relieve employers not just of the hard work of forecasting production flows and targets, but of the two or three people doing the book-keeping.

The figures that illustrate the spread of that revolution are awesome: in 1960 there were 500 computers in operation. By 1970 this had reached 6,500; by 1980, 175,000; and by 1983, 750,000. For smaller business computers the figures are more startling still. There is a rather whimsical belief in some government circles that as the machines come into operation, they have not destroyed jobs, merely added new ones. Yet the number of jobs in high technology industries since 1975 has actually fallen from 150,000 to 120,000 in 1983, even though the industry has been growing by 12 per cent a year. Overall, some 1.8 million manufacturing jobs were lost in 1979–84, and only 132,000 jobs created. The large majority of jobs have been lost through the shrinkage of markets for traditional industries. But an increasing proportion has been lost directly through the introduction of the new technology, although it is impossible to estimate exactly how many. The new technology has created many part-time jobs: of the 613,000 new jobs created since early 1983, 386,000 were part-time; full-time jobs fell by 129,000 during the period.

The figure has the potential to be awesomely higher: a Swedish computer system could effectively take over the running of the country's mines, leaving a skeleton maintenance staff in each pit. A major robot manufacturer told the authors that the application of assembly line robots was in its infancy. Another wan hope often expressed is that the technological revolution will look after itself: as the new technology creates new demand, so industries feeding paper at each other – new types of accountancy, of legal and advice services, for example – will provide the employment. The hope is a

1

slight one: computers (whatever hundreds of frustrated users may say) simplify work, and cut through bureaucratic procedures. Similarly, leisure activities, although booming, are expanding at nothing like the kind of job replacement rates needed to cope with the computer revolution.

Politicians have so far prefered to gaze into the deep blue yonder rather than examine the implications of this revolution. Few understand computers, still fewer have any answers to the staggering impact upon working practices of the new technological revolution. The direct effect upon manufacturing jobs has already been mentioned. There is virtually no industry with tasks of a repetitive or mechanical kind that cannot be operated through the new technology: the lack of new investment and the resistance of Britain's unions are all that stand between many workers and the dole queue. Most sub-managerial office work is repetitive and computerisable; here the reluctance of employers and the high cost of shedding labour, rather than union strength, combines with a natural resistance to innovation to prevent the new technology being introduced.

In the public sector – in particular in the clerical, tax-collecting and administrative grades – the scope for manpower savings is enormous. Private service industries are the ones most under pressure at the moment. As innovative firms begin to use the new technology to greater effect, their overmanned competitors are being driven out of business. It will soon become vital for most medium-size service sector industries to bring in the new technology merely in order to keep up.

The most enlightened companies benefiting from the new technology are using it not just to reduce their staff and increase productivity, but to alter the working practices of their employees: word-processors can be operated from home, and a very few employees are being allowed to do just this. Flexi-hours are being introduced, allowing workers to be employed at reasonable rates of pay for shorter hours. Sabbaticals are becoming more common, allowing workers to take a year off every ten years of their working lives. The growth of part-time work has been spectacular: in 1951 there were 800,000 part-timers, 4 per cent of the workforce; today there are 5 million, 20 per cent of the workforce, 80 per cent of them women. Nearly 1 million of these are in the retail trade, some 800,000 in education, 600,000 in catering and 500,000 in nursing. We believe this should be massively extended, particularly in the public sector, and the rights of part-timers should be defined.

New systems of shift work are being introduced into some companies, with three-day weekends and shorter hours in others. Most of the new arrangements are the exception rather than the rule, however: they challenge traditional concepts of industrial discipline. Most motivated employees prefer to work a full week and even overtime; and employers are usually reluctant to use the new technology to ease the load on their employees. A company's main purpose, quite rightly, is to take advantage of the savings involved to maximise profits.

The profit motive is one of the great driving forces of human society, and needs to be harnessed to pull society forward, not to be stopped or allowed to run away, leaving society stranded. The Labour Party, and to some extent the SLD, favour standing in the way of the inevitable. Conservatives believe that the task at the moment is to allow progress, spurred by the new technology, to move forward in the service of society. Parliament, which by and large prefers to talk about an earlier form of industrial organisation, has not managed to get to grips with the implications of the new technology. The dizzying opportunities for a more prosperous, more leisured, freer society offered by the new technology are being ignored. Worse, no effort is being made to cope with the new forms of inequality (which transcend the old political debate) being created by the new technology.

The new inequality is that of a growing number of people unable to take advantage of the opportunities created by the new technology. They are, first, manual and white-collar workers in their middle to late working life, too old to be retrained to use the new technology, and for whom there are no jobs anyway. Because of the extent of unionisation and employment protection legislation, the major problem, however, is at the other end of the job market: untrained school-leavers and even MSC-trained workers for whom too few opportunities are being created, as the new technology replaces the old. One million of the 2.3 million unemployed have been out of work now for more than a year. Some 750,000 have reached the age of 25 without ever having had a job.

It is absurd to argue that the level of supplementary benefit is an incentive to stay on the dole. There are some malingerers, yes; but only very low-paid jobs have wage rates comparable with dole or social security payments of, for a single person, £27 a week. The average industrial wage is around 2½ times what can be earned on social security, and offers major fringe benefits. Moreover, the value of social security benefits, which is tied to inflation, has been growing more slowly than average industrial and white-collar wages, which have grown faster than inflation. It is reckoned that only 6 per cent of unemployed men are better off than in work (because they have large families). Nearly 50 per cent earn less than half their previous income in work. Above the average earnings sector, the new technology is creating an upper class of jobs and skills that are considerably better paid than the average white- or blue-collar worker. On top of that, the profits generated by those installing new technology allow for (a) a higher shareholder return; (b) higher management remuneration; (c) the ploughing back of investment into wealth-creating, capital-intensive expansion, which again does not create many new jobs.

Most of these trends are admirable and, this book will argue, must not be obstructed by blind anti-innovation prejudice on the part of the workforce. But such responses are inevitable if the government pays no attention to the new realities. These are, on the one hand, that a small minority is doing very well out of the new technological revolution; that a large majority is doing quite well; that Britain as a whole is prospering because of it – indeed, would fall behind other nations unless it continued to innovate. Yet, on the other hand, there is a substantial and growing minority – if families are counted, as many as one in five of the people of Britain – condemned to no prospect of permanent employment; to life on an adequate but far from comfortable income; to idleness; to no prospects or hopes of betterment; to no hope of home ownership; to no stake in the capital-owning democracy the Conservative Party is committed to bringing into being.

This book argues, firstly, that it is technological innovation that will contribute – if its effect upon working practices is unregulated by politicians – to the growth of long-term unemployment, whatever the prospects for the rest of the economy. Secondly, the expansion of this core of unemployed is continuing to add to all of Britain's social problems – poor housing, education, inner-city decay, crime, racial conflict, and ultimately, possibly social violence; those problems will be ameliorated to the extent that this is tackled. Thirdly, it is intolerable in a developed country in the last quarter of the twentieth century for this ghetto to be allowed to increase. It has been the Conservative Party's historic responsibility to ease the effects of uncaring progress, to safeguard the wider interests of society, initially through the revolution of agrarian enclosures and then the Industrial Revolution. It is the Conservative Party's responsibility to see that the technological revolution brings benefits to all the people of this country, not just the fortunate majority. Only then can Britain's unique social order remain in peace, and its values be preserved.

The Conservative Party is much better placed than other political forces to perform that task. Conservatism is not a dogmatic creed; but it is not true that it has no principles either. It seeks to apply unchanging principles to change, which it accepts as both desirable and inevitable. Those unchanging principles can be briefly summarised: firstly, that change should benefit the greatest number; secondly, that change should enhance the freedom of the individual to do as he or she pleases without prejudice to the freedom of others; thirdly, that the most successful and strong in society have a social responsibility towards those less successful and weaker. Conservatives believe that any attempt to obstruct progress is doomed, and can only, in the long-term, result in a poorer, more inward-looking society; but progress which pays no regard to these principles is dangerous and can wreak economic, social and, ultimately, political havoc.

Take two extreme examples outside the economic sphere. The development of nuclear science and biotechnology cannot be stopped. Yet if both are developed without political regulation to protect society's interests, the effects are potentially disastrous. Politicians have wrestled with only quarter success to keep nuclear technology as a boon, not a disaster, for mankind; tampering with genes remains alarmingly out of their control.

Rooted in these principles, Conservatism is pragmatic. Anything that furthers these principles is acceptable. Different policies are suitable at different times. A greater degree of state intervention on behalf of the worse off was aceptable in Disraeli's time and during the 1930s because change was obviously failing to benefit a lot of people. Similarly, Conservatives accepted the beneficial part of the welfare state, while shedding the dross, during the 1950s. When economic growth ground to a halt in the mid-1960s, Conservative minds were quite rightly concentrated on how to renew economic growth, rather than on how to redistribute income more equitably. This manifested itself in the philosophy evolved at the Selsdon Park conference under the leadership of Edward Heath.

In 1972, after two years in government, his experience suggested that economic expansion combined with wage restraint offered the best way forward. His government was tripped up by the miners' strike before the effects of his policies could be fully felt. After all

possible benefits had been dissipated by the 1974–9 Labour government, Mrs Thatcher's government took office pledged to monetarist
policies in order to keep inflation under control and provide the
climate for business confidence that had so conspicuously failed to
materialise in the preceding ten years. The climate was created, at a
cost, and was taken advantage of, creating six years of good economic
recovery.

The point about these policies is that none of them are articles of
faith. They are there to be tried, tested, discarded if they fail,
improved upon. But Conservatives are not wedded to monetarism, to
looser credit, incomes policies, fiscal incentives or balanced budgets
as articles of faith. These are but means to the ends of increasing and
spreading prosperity, enlarging the freedom of the individual, and
social obligation, which are all that Conservatism aspires towards.

'What do you want to do with your power?', a journalist recently
asked a Conservative MP. He could not understand the response:
'Govern efficiently.' The Conservative Party is a managerial party, in
the continental phrase, a technocratic party. It wants to make things
work, and is prepared to consider any suggestion. Its commitment to
social stability means that it wants to make things work for the benefit
of everyone, and gives it its concept of social duty.

Conservative pragmatism makes it uniquely fitted to cope with the
challenges of the technological revolution. The other two major
parties are rooted in ideologies embedded in problems of the past,
which are losing their relevance to the future. Take the Labour Party
first: it is the offspring of the trade union movement, its philosophy
rooted not in Marxism but in a British tradition of, on the one hand,
the Fabian intellectual socialism behind the ideal of the welfare state,
and, on the other, straightforward worker self-betterment through
industrial action. This trade union strand, although lightly diluted as
the twentieth century proceeded, remained constant: its thinking
evolved from courageous self-assertion in the face of considerable
odds at the end of the nineteenth and beginning of the twentieth
centuries, to increasing bloody-mindedness just before and after the
First World War, to timid introspection after the strikes of the 1930s
and the failure of the General Strike in 1926, to sturdy patriotism in
the 1930s and 1940s and most of the 1950s, to a reassertion of
self-interest at the expense of the rest of society and of each other in
the 1960s and 1970s.

The end of the 1970s and the beginning of the 1980s has seen a
slight chastening in union attitudes, as a result of mass unemployment

and their own unpopularity. With the exception of the late 1960s and 1970s, however, the ideological content of union thinking – and hence the thinking of a large part of the Labour party – has been quite small. On the intellectual side of the Labour Party, however, a considerable streak of utopian socialism was injected alongside Fabian theory at the time when all was thought to be rosy in the socialist haven of Stalin's Russia.

After the Second World War, when this was discovered to be far from the truth, part of the Marxist left surprisingly became more entrenched in its beliefs, squabbling with Trotskyists who considered that socialism had never really been tried in Russia. These groups have gained considerable support in the constituency parties. The majority of the Labour Party, however, followed the view best articulated by Anthony Crosland, which elevated the welfare state to a pinnacle.

All these elements of Labour thinking run into trouble when faced by practical decision-making and, more importantly, the technological revolution. Take the trade union heart of the party. Its two, non-ideological, priorities are the maintenance of jobs and the maximisation of wages. This makes the unions fairly trenchantly opposed to technological innovation, which takes jobs away and alters working patterns so that wage rates, in order to spread work more evenly, may have to be reduced. The argument that the maintenance of jobs in industries which have become uncompetitive through the lack of technological innovation in the end leads to greater dislocation and job losses has taken a long time to percolate through.

Take the welfare-state Croslandite heart of the party: their belief was that economic growth would in theory favour both greater prosperity for all, and a state which would provide ever greater benefits to those lower down the scale. When economic growth stopped, the Croslandites never acknowledged that excessive state spending itself might have acted as a brake on growth. The Fabian part of the party was left floundering for ideas as to how it might be started again: a government-stimulated public works programme seemed the only option. This view failed to take into account (a) that the public sector was sluggard and inefficient by comparison with the private sector and hence a very slow locomotive for growth; (b) that any injection of demand into the economy tended to be dissipated on imports anyway; (c) that state intervention, to which the Fabians were committed, tended to be concentrated in precisely those traditional industries which were likely to suffer most from rationalisation and the introduction of the new technology. Welfare state socialists viewed the new technology with apprehension.

Finally, the Marxist wing of the Labour Party, which was becoming disproportionately powerful in the constituency parties, persisted in seeing the political struggle in class terms, as between exploiters and exploited, with the state's role to act on behalf of the latter. They were quite oblivious of the fact that the middle class had grown considerably by the mid-1960s; that a large part of the working class had become middle class in taste; that even in socialist societies the state itself had become the political expression of the middle classes; and of the abundant evidence of the inefficiency of state control. The result was that the Marxist wing's leading spokesman, Tony Benn, has next to nothing to say on the technological revolution (although he himself is a former Minister of Technology). The battle between the classes was being transcended by the battle between the ever more prosperous employed, and the ever greater minority of unemployed.

The thinking of the Social Democratic Party rather faithfully descends from the Croslandite tradition, which partly explained the party's extraordinarily cautious approach (especially in view of it own claim to be innovative) to the subject of the new technology. Broadly speaking the party welcomed the advent of the technological revolution and of its impact in revolutionising working hours; but the problem of unemployment was, in the Social Democrats' view, to be solved by deficit spending. The Liberals make very little allusion in their literature to the technological revolution, despite its potential contribution to their utopia of restoring Britain to the green and pleasant land of old. (The Liberal Party, which historically was the party of industrial capital, had of course played little part in introducing social legislation to soften the impact of social change; but as the party has been out of government for so long, it is a little unfair to saddle it with its historical sins.) It is strange, nevertheless, that a party which is in many ways a laboratory of new ideas should be so hesitant on this issue.

Nevertheless, the subject is likely eventually to force its way to the attention of the other parties. Before then, the Conservatives, as a party unafraid of change, that seeks to mould change in the best interests of the British people, should launch its revolution in working practices, so as to redistribute work from those who seek less to those who seek more.

Employment should be at the very top of the political agenda.

Inflation and stagnation have receded as problems, thanks to the remarkable success of Conservative policies in recent years. That success is undeniable. Two problems had bedevilled British growth since the mid-1960s: first, the growing cost of labour, at the expense of profits and investment; and second, the growth of government spending. Although the origins of the second problem go back to the 1950s, the problem of excess wages began under the Wilson government of 1964–70. The Heath government unsuccessfully made a dash for growth through greater public spending, but its chances of success were dissipated by the defeat of its incomes policy during the miners' strike of 1974. After 1975 the Wilson government exacerbated both problems, and the Callaghan government spent the remainder of the Labour term under IMF pressure trying to coax the two genies back into the bottle.

The approach of Mrs Thatcher's government was to allow a modest wage explosion to begin with; this was eventually controlled through restricting credit and the money supply which, by forcing companies to shed surplus labour and cut down on wage bills, worked more effectively, but at higher cost, than an incomes policy. An incomes policy would probably not have worked, against the prevailing triumphalism in the union movement at the end of the 1970s. At the same time, recession and the higher cost of social security pushed up government spending, which only began to fall as a percentage of GDP at the beginning of the 1980s.

A reduction in inflation has contributed to a climate in which business investment is perking up, but not to the degree necessary to replace lost jobs in manufacturing industry. Meanwhile an effective incomes policy is coping with public sector wage demands: private sector wages are being determined by the market. Earnings, however, continue to rise faster than inflation, but not excessively so. The danger on the wages front is that any pickup in demand could be the spur for fresh militancy on wages; high unemployment may have a moderating effect, but the government may have to brace itself for wage confrontation – in which it will have to stand firm or lose the recovery – over the next few years. The danger of government spending getting out of control is probably past, unless unemployment (and hence the social security bill) soars. The reduction in state spending is allowing major tax cuts to be made.

The impact of the new technology on this picture will be considerable. If the new technology is introduced without adequate regulation, unions in many industries are liable to secure better wages for full-timers working in them. The danger is that companies will be prompted to lay off workers – or simply recruit fewer – rather than challenge the unions over wages. This will contribute to high unemployment, which in turn will endanger the government's public spending targets. The worst of all worlds would be for inflation to start to increase as a result of demand-generating wage pressures, while unemployment and government spending simultaneously rise.

Chapter 2 details the effect of the technological revolution on the wider economy. In Chapter 3, Peter Walker looks at how small amounts of public spending can catalyse employment creation. In Chapter 4 Michael Heseltine looks at the regional aspects of the new unemployment. In Chapter 5, Jim Lester and others look at the problem of youth unemployment and the need for a skills improvement programme. Chapter 6 looks at the need for infrastructure improvement. In Chapter 7, the European dimension is examined, while Chapter 8 looks at Britain's role in relation to the developing world. Chapter 9 argues that, in a country of limited natural resources and space like Britain, and in a society in which leisure will play an ever larger role, protection of the environment is not a luxury, but a necessity; more far-reaching and comprehensive environmental legislation is required. Chapter 10 argues that society has dangerously passed by our present constitutional set-up: Parliament is in long-term danger of becoming a rubber stamp, approving the decisions of the executive because of its failure to reform itself to meet the mushrooming demands of modern society. The chapter suggests ways in which democracy, which had been in constant evolution in Britain until the 1930s, might be deepened.

The proposals in this book are radical, but they are Conservative. Unless society imposes its own order upon the direction of change, society will be changed much more radically, and in ways that could be dangerous. The values that Conservatives believe in are as relevant today as they ever were, and should be reasserted. Conservatives have always had a genius for regulating change, and shaping change in Britain's interests.

Take Disraeli's attack upon the Liberals, the cruel, callous, unthinking progressives of his day:

> Liberal opinions are the opinions of those who would be free from certain restraints and regulations, from a certain dependence and duty which are deemed necessary for the general or popular welfare. Liberal opinions are very convenient opinions for the rich

and powerful. They ensure enjoyment and are opposed to self-sacrifice. The holder of liberal opinions maintains, for example, that the possession of land is to be considered in a commercial light and no other.

Or take Lord Salisbury's view: 'No system that is not just as between the rich and the poor can hope to survive.' Stanley Baldwin's objective was to 'elevate the condition of the people'. Churchill believed in crude commonsense economics: 'The community lacks goods and a million and a quarter people lack work. It is certainly one of the highest functions of national finance and credit to bridge the gap between the two.' Harold Macmillan argued that 'The important thing is that society should be organised in such a way as to bring the economic system under conscious direction and control, and that the increased production should be directed towards raising the standard of comfort and security for all the people.'

Iain Macleod held that 'I believe we should seek to create a nation increasingly efficient, increasingly tolerant, and, above all, increasingly just.' Edward Heath was passionately committed to 'the middle way'. Sir Keith Joseph argued that 'monetarism is not enough'. Mrs Thatcher has said bluntly: 'of course we care about unemployment'.

The new technology is ushering in a new society, and the new politics must not be far behind. It is within the proudest traditions of the Conservative Party to rise to the challenge.

2 The New Unemployment
Robert Harvey

A recent OECD report underlined the awesome problem of unemployment in Europe: in spite of the continent's economic recovery, unemployment increased from 19 million in 1984 in the OECD countries outside the United States to 19.5 million in mid-1985. In the four major Western European countries – Britain, France, West Germany and Italy – unemployment for young people has risen from 21 per cent in 1983 to 22 per cent in 1984 to nearly 23 per cent in 1985. The report concludes:

> In short, contrary to all past experience with youth unemployment in the OECD area, the current economic recovery is not expected to yield appreciable dividends in increased employment opportunities for young people in most OECD countries, or reduced unemployment rates.

The report pinpoints four other ways in which the present unemployment crisis differs from previous ones: unemployment for disadvantaged youth and the disabled has risen; paradoxically, unemployment among the well educated has also risen faster than the norm; labour force participation rates – the percentage of people in work – have continued to go down; and young people in their twenties with no working experience have a diminishing chance of ever getting a job – which leads the OECD to define them as a 'high-risk generation'. In West Germany and the USA youth unemployment has declined only slowly, in spite of vigorous economic recoveries: in West Germany the rate of youth unemployment is expected to fall from 11 per cent in 1983 to 9 per cent in 1986; in the United States from 16 per cent to a still-high 12.5 per cent. In Canada the rate is expected to fall from 20 per cent to 17 per cent, and in Britain just from 22.5 per cent to 21 per cent. Elsewhere it is actually going up: in France from 21 per cent in 1983 to 31 per cent in 1986; in Italy from 32 per cent to 37 per cent; in Spain from 39 per cent to 46.5 per cent.

The youth crisis is one disturbing aspect of the new unemployment. Another is the emergence, in Britain at least, of two virtually parallel economies – a booming service-based one in the South East and a declining manufacturing one elsewhere. Between 1979 and 1984, manufacturing as a percentage of GDP shrank by 2.4 per cent and construction by 0.5 per cent, while services grew by 4.5 per cent, energy and water by 0.9 per cent and agriculture by 0.6 per cent. In the decade to 1984 services grew by an amazing 22 per cent of GDP in real terms – 7 per cent more than the economy as a whole (financial services grew by 70 per cent). Over the decade some 2.7 million jobs were lost in manufacturing, and only around 1.2 million jobs were created in services. Only a few manufacturing industries performed well: electrical engineering grew by 23 per cent over the decade, and chemicals by 13 per cent. Mechanical engineering slumped by 30 per cent, motor manufactures by 36 per cent and textiles by 31 per cent. Between 1979 and 1984 only 132,000 service jobs were created, while 1.8 million manufacturing jobs were lost.

Set against this depressing scenario for overall job creation there are two schools of thought about the impact of new technology on employment. One argues passionately that the new technology is taking away existing jobs, leading to a smaller and smaller concentration of work and wealth in the hands of fewer people. The other argues that as the new technology takes away existing jobs in manufacturing industries, the new wealth will generate a new demand for services, which will take up the slack. The first school urges resistance to technological change, the second embraces it with open arms.

We, like the second school, unequivocally embrace the new technology. It is inevitable, in the sense that if a competitor nation has it and we do not, our industries will be unable to survive except behind closed tariff walls as an out-of-date, offshore economy. There is no choice in the matter: Britain has never shirked innovation, and if we are up there with the world leaders, the country will be all the richer for it.

But we do not share the optimism about jobs of the second school. In time, service industries may lead to a full employment society. In the medium term, many people have lost and will continue to lose their jobs in newly automated manufacturing industries, and will be unemployable in the new skills. As has been argued in the previous chapter, the answer is not just to train them in the new skills, but radically to reshape the concept of work, so that early retirement, job-sharing and reduced hours of work become the norm rather than the exception.

What is the evidence for this? Much pioneering work has been done by Professor Christopher Freeman of the University of Sussex, and Luc Soete of Stanford University. Both are academically cautious in their conclusion, which is nevertheless a dramatic one: that the introduction of information technology amounts to a 'paradigm change' (economese for a new industrial revolution) which will sharply alter the organisation of work. Professor Freeman and Mr Soete argue in *Information Technology and Employment* (University of Sussex Science Policy Research Unit paper, 1985) that there are two separate components of the technological revolution: the introduction of robotics into assembly-line manufacturing, and the creation of new products and processes; and the information technology revolution itself, which allows much greater and more diffuse production-line flexibility.

In addition to the impact on employment, the technological revolution will lead to a drastic reduction in the 'costs of many products and services, provoking widespread opposition from new professional interests'; and the new technology will also allow

> scope for an entirely new range of products and services and for a dramatic improvement in the technical characteristics of many other products and processes – for example increased reliability, accuracy, speed etc. The benefits of this cannot be exaggerated: a much greater range of goods, of much higher quality at a much cheaper price can be produced, allowing a much more varied society than exists at present. In such a society smaller production units are likely to be found, and smaller businesses create employment, as has been the experience in the United States. This will be a great leap forward. The only sufferers are those whose skills and lack of education and initiative ill-equip them for this kind of work.

The figures, for Britain at least, make a pretty convincing case: although the information technology industry itself has been growing by around 12 per cent a year, employment within it has actually fallen from 150,000 in 1975 to 120,000 in 1983. Studies show that 60 per cent of firms introducing computers have done so to cut back on jobs.

Employment in traditional manufacturing industries has fallen dramatically between 1971 and 1981 – by 24 per cent (200,000) among skilled operators, by 21 per cent among ordinary operatives (1 million). In lower grade service industries there has been no corresponding increase: the number of those in clerical jobs has fallen by around 2 per cent (50,000), although the number in security jobs has surprisingly increased by 5 per cent (40,000); there has been a 5 per cent (50,000) fall in the number of salesmen, along with a jump of 39 per cent (50,000) in the number of supervisors.

The number of engineers and scientists has, not surprisingly, increased by 28 per cent (30,000) and the number of technicians by 6 per cent (30,000). The biggest increase has been registered in those service areas where there are fewer jobs available: quality services, such as education, health and welfare, the professions, and literary, artistic and sports activities. Employment in education-related activities, for example, is up by 160,000, a 23 per cent increase; in health and welfare it is up by 300,000 or 46 per cent (both of these, however, have benefited from government support); employment in the professions went up 150,000 (20 per cent) and in literary, artistic and sports activities by 28,000 (17 per cent).

An authoritative report by the Manpower Services Commission last year (1987) looked at the impact of the new technology in various sectors: the impact on the shop floor, concludes the MSC, will mainly be felt in three areas. The first is non-complex assembly machines, which are taking over the sorting, packing and flow production lines: 'all these forms of automatic handling mechanised low skill repetitive assembly operations, tasks which would otherwise have been done by semi-skilled workers'. The second area is automatic insertion – the automatic placing of electronic components into printed circuit boards: 'The widespread diffusion of this is just beginning and its advantages are large, in terms of its flexibility through reprogramming and the facility to combine assembly with testing at all stages of production. It is much faster than manual assembly, and is less prone to error.' Third is assembly-line robots, which are in their infancy. In 1983 Britain only had some 1,750 robots, of which only 6 per cent were in assembly lines. Yet IBM, for one, believes that the introduction of robots will become widespread. Skilled labour is likely to be the chief casualty of industrial robots. The MSC report concludes: 'when sophisticated robots become common, a much greater proportion of industrial jobs will be susceptible to automation.'

In the service sector, the MSC identifies typists, secretaries and administrative staff as those most at risk. Between 1981 and 1983 the number of clerks fell by 6 per cent; during the same period the number of typists and secretaries fell by 7 per cent. Some studies suggest that over the next ten years, some 20 per cent to 40 per cent of secretarial jobs could be lost, rising to 70 per cent in the long term. In many offices secretaries are already little more than adjuncts to the boss's ego. The MSC study is not particularly optimistic about the growth of jobs in any but the top professional and high quality services, like banking.

What has been the experience in other countries? The United States is often held up as an example of a country where the number of jobs has surged since the introduction of new technology: in practice, many of the jobs are connected to government services, and many others to dynamic small-enterprise firms for which America has a tradition of venture capital. Some studies are very pessimistic: a recent report by W. G. McDerment of CEDEFOP, the European Centre for the Development of Vocational Training, points out that 'it is difficult to go into a large car making plant in the United States and see the very small number of people working in the big assembly shops since the introduction of large numbers of robots to believe that the impact on employment will be negligible' (*New Technologies, Education and Vocational Training in the USA*). One manufacturer acknowledged to him in the same report that 'Many companies will be forced to reduce the number of people employed.' Dr Richard Darf, of the University of California, said that '80% of installed robots are used within the automobile industry only for ... painting, welding and materials handling.'

The US robot industry is expected to grow rapidly. The total number of robots installed in the USA rose from 600 to 3000 between 1975 and 1980 and is expected to reach an estimated 125,000 by 1990. With regard to labour displacement, Dr Dorf's personal conclusion was that it would become a crucial issue 'if robots and automatic manufacturing are rapidly applied.' A major study by Wassily Leontief concludes that the introduction of information technology would make it possible to save up to 20 million workers, or 12 per cent of the American workforce by the year 2000.

In Canada a report put out by the Canadian Employment and Immigration Commission suggests that out of thirty-nine labour sectors, only eight would benefit in terms of jobs from the introduction of new technology. Jobs would be lost in all the others: if industry employed up-to-date technology, the study concludes that some 600,000 jobs would be saved, although in the long term the study believes that a greater number would be created. The Canadian Department of Employment, in another study, baldly acknowledges that there will be 'a decline in the demand for unskilled occupations in manufacturing and increasingly in clerical occupations in the service sector.'

Japanese studies suggest that the introduction of further automation in what is already a highly automated economy would on the whole be beneficial to jobs because it would help to promote exports. However, a major study acknowledges that the number of jobs created in employment throughout industry will only be around half of those displaced by the new processes; a growth in youth unemployment (around 6 per cent compared with 2.5 per cent overall) is worrying the Japanese.

In Germany a report drawn up by Siemens suggests that 25 per cent of all office jobs could be completely automated. A reduction in working hours and re-education have been in the forefront of the German response to the problem – these are the measures recommended elsewhere in this book as the quickest solutions to structural unemployment.

Generalised reflation would merely make it possible for the most profitable parts of the economy to modernise faster (but not provide more jobs) while undermining external confidence in the economy and causing inflation. The tight-money approach simply does not address itself to the problem of unemployment at all. The answer we favour is selective, but conservative and non-inflationary, government intervention to provide jobs and benefits for those unlikely to be employed again as a result of the technological revolution; and long-term government action to adapt working patterns to fit in with a society in which there is less work to go around, so that everyone can share in its greater wealth.

The idea that the Conservative Party is inherently opposed to state intervention is wrong. It is positively in favour of state intervention where this is necessary: for example, Conservatives support the police and the armed forces because of the fear of what would ensue if criminals and some foreign governments were left unchecked. Conservatives believe that the less state intervention in the economy the better, because practice has shown that more wealth is created if people are left to get on with it. But this is not a matter of ideology. Conservatives support state intervention where this is necessary to prevent, for example, the exploitation of the workforce in the process of creating wealth (Conservatives were the prime movers behind nineteenth- century social reform); or the worsening of living standards for those least able to help themselves; or the isolation of one section – such as the unemployed – from the community.

So with the effect upon employment of the new technology. Conservatives start from the premise that the introduction of the new technology is of great benefit, because it means that more wealth can be created using less sweated labour in monotonous jobs than the old industries, and at less cost to the environment. But that does not mean that the state has no role to play in harnessing the effects of the new technology to the benefit to the whole community. The response of some Conservatives to the new technology is to insist, as an article of faith, that it will create as many jobs as it replaces: this is vigorously argued, for example, in *No End of Jobs*, by Michael Marshall, Charles Clark and Philip Virgo:

> In the majority of concerns, improved levels of service are far more important than mere headcount reductions. Free from form-filling and bureaucracy, the staff can give more attention to the custo-mers' needs, and this means more business. Any reduced manning levels can often be used to allow staff to work shorter hours in return for producing more service where the customer wants it – after 5 pm and on Saturday and Sunday (not just in the private sector but tax and local government offices, nationalised industry, energy and services firms and so on).

The pamphlet goes on to argue that jobs should abound in the areas of domestic service and of traditional crafts.

The argument begs a number of key points. First it applies the lesson of a few types of industry across the board to the rest of industry. Second, already established firms do react to the new technology by keeping on staff and trying to improve quality, or to bring in new business. But many such firms are merely rather conservative industries failing to benefit from the full labour – and hence cost-saving – advantages of the new technology, and hardly therefore the harbingers of a new industrial future. The new firms coming on to the market mainly to maximise the labour-saving effects of the new technology are unlikely to be so wasteful. Product improvement and the saving of labour are both possible under the new technology. As for domestic service and crafts, it has long been true that there has been a mystifying shortage of people prepared to take this on. But this is a result of rising expectations, and short of retaining a pool of unemployed people that have to take such jobs, it is hard to see how the new technology will help. The new technology will still create some jobs and lose many more. A case-by-case look at industry illustrates this point.

In *agriculture*, the most mechanised of British industries, most of the jobs that could have been lost through the introduction of new technology have been. There is no prospect of new jobs being created in this field.

In *basic manufacturing industries*, some 1.8 million jobs have been lost in the past ten years. Two factors have contributed: one has been the introduction of automation along assembly lines, the second a shrinking in Britain's traditional markets, domestic and foreign, as the country grows less competitive due to high labour costs and low investment. The recent shakeouts in major industries have made British industry more competitive, and have resulted in the recapture of some lost markets; but part of the price has been the introduction of new labour-saving technology. Greater quality control and product improvement has made the decline less than it would otherwise have been. All the same, the authors could find no major forecasters who would predict anything other than a continuing decline in the number of those engaged in manufacturing.

In smaller *ancillary firms supporting manufacturing companies*, and in *services associated with manufacturing industries*, such as motor suppliers and garages, prospects for employment have seemed unlikely to improve. Demand in the first area is associated with consumption by manufacturing (which is declining) and in demand for those services, which is likely to grow as general economic prosperity improves. As the new technology contributes to greater prosperity, so the employed working classes and salaried middle classes should have more money to buy better cars, which require improved servicing. But the second category is in the hands of small firms which are unlikely to shed labour as a result of the introduction of the new technology, but which are favourites for growth in this field.

In *domestic service*, the prospects for growth seem slighter. Domestic servants' wages remain too high for the bulk of families to be able to pay them because the pool of available labour in this area is a small one. There has been a growth in the field of care for old people etc., but the labour-saving devices introduced in this field, coupled with the growing cost of sending people to homes, means that the number of jobs in this sector is limited, and confined to the care of the minority of the middle class able to afford such services.

Information technology and *financial and legal services* are the biggest growth areas for new jobs. But some optimistic forecasts need to be looked at carefully. It has been reliably reported that some 50,000 new jobs could be created by the laying of cable television lines; the jobs are temporary, however, and based on a rosy idea of how many British households will switch on – assumptions which are clearly being drastically revised downwards. Video production and entertainment could provide many new jobs – maybe as many as 100,000. Some 200,000–300,000 could be provided by the provision of new learning materials, assuming that the government is prepared to provide education authorities with the money to supplement the rather small private sector demand for these products. Software production and information technology products themselves are likely to be a boom area, providing as many as 150,000 new jobs.

The field of legal, accountancy and professional services requires two conditions for new jobs to be created: first, a rise in the demand for them, as the prosperity engendered by the new technology creates new needs; second, a lowering of the cost of these services as a result of the introduction of the new technology. The trend, however, – and the admirable ambition of Conservative governments in recent years – has been to simplify the tax and legal systems, so reducing the demand for such people. And any new technology-related cost reductions are likely to result in fewer jobs in those industries, not more. Among office and clerical staff, there is likely to be a dramatic reduction in jobs as a result of the introduction of the new technology. Most routine secretarial and ledger clerk jobs are directly replaceable by quite primitive computers; and if, as seems likely, typewriters that can be dictated to become economically viable, the sector seems likely to contract further.

The other area often pinpointed as a growth one is that of *small services*, catering for such needs as alternative technology and local newspapers. These are necessarily likely to be small employers, and unlikely to have any significant effect on the jobs market.

Yet if the overall picture is likely to be one of lower job prospects, what is to be done? Clearly the remedies of the Labour and Liberal Democratic parties, simply to expand public works projects and subsidise industries that might otherwise close their gates, is merely an escape, and one likely to lead to worse trouble in the long run. The Conservative approach should be one of encouraging the new industries, while spreading the benefits they create. Specifically, Conservatives acknowledge that for growing numbers of young people the idea of a forty-hour week until retirement age is not an ideal way of life, while for many older workers the idea of greater leisure before retirement is an attractive one. Greater freedom of choice should mean greater freedom to diversify work, within economic and organisational limits.

The public sector needs to give a lead. The job-splitting scheme as conceived by the Department of Employment has proved a limited success so far because it applies to the private sector, and it has run up against the natural reluctance of managers to complicate their task without visible reward, and of unions to allow practices which might undercut wages. In the public sector the state is the employer and, except in the case of a few nationalised industries, is in a position to dictate to the unions. The simplest recommendation would be that two jobs be created for every retiring public sector employee, or for every employee who volunteered to share his or her job, at the same overall cost as the job being split. The eventual breakdown might be an eighteen-hour week for two people, working six hours for three days each. National insurance contributions could be modified accordingly, as would other employee benefits. State retirement pensions would be unaffected. It must be stressed that no one in an existing civil service job who wanted to work a full week would be obliged to share jobs, and there would still be jobs available to those who wanted to work a full week as new entrants to the labour market. A part-timer could switch to a full week if he or she wanted, as so many married men with families seeking to maximise their earnings would want to. But many new entrants, and many workers aged 40 or more, would welcome the chance to work less than the full week.

The administrative costs of such a scheme are not great, involving usually no more than the training of the larger workforce that would be required, and a competent office administration to ensure that desks were properly manned. It is reckoned that, within the spread of three years needed to introduce such a scheme, as many as one million split-jobs could be created in the civil service alone, and as many as 500,000 in the nationalised industries.

Union resistance, of course, would be virulent and considerable. The change would be seen as undermining the concept of a full day's wage for a full day's work. Yet even the TUC has begun to accept the need for a dialogue to absorb the full implications of the new technology. Presented as a job-creating package, it would be hard for civil service unions to mount an all-out strike, and the Institution of Professional Civil Servants and the CPSA are traditionally moderate bodies. Provided there was no element of compulsion – job-sharing would apply only to those who wanted it – union opposition might be limited. This single measure, we believe, would have a major impact on reducing unemployment and changing traditional attitudes to work and working hours. The eighteen-hour week could be flexible, and extended to evenings and Sundays. These are times when many people require local government services, for example, but cannot get them, because they are themselves at work. The example of the public sector would, hopefully, be followed by the private sector.

We do not believe that the private sector should be bludgeoned into accepting schemes to reduce unemployment: the wealth-creating part of the economy is best left to its own devices. All the same, a number of fiscal incentives can be made available for firms prepared to initiate job-sharing. The government, as suggested in an excellent Bow Paper, *Micro Solutions to Unemployment*, could offer to make up company pension funds for employees who opt for job sharing, so that long-term pension prospects are not affected. Employers who increase capital costs in launching job-splitting schemes could receive favourable tax treatment. In the small business sector part-time employees and job-splitting will be favoured by a decision to raise the thresholds for paying VAT to some £60,000 and by allowing employers wider latitude to escape the restrictions of employment protection legislation for workers earning below a certain threshold.

Hand in hand with job-sharing goes a more flexible approach to working hours. The initial costs of new working arrangements would be considerable, which is why few employers will risk it. The government could investigate the possibility of a reduction in the working week. This is the solution favoured by the trade unions, who insist, however, that there must be no reduction in wages. In most companies this would simply entail considerable additional costs and no savings. For this reason we oppose a general reduction in the working week.

We do, however, urge industrial workers and companies to put this into effect wherever the new technology makes a shorter working week practicable and wherever workforces are willing to work fewer hours for less pay. Again, the nationalised industries can give a lead. It is often said that experience shows that workers will work longer hours for more, but never shorter hours for less: but very major differences of motivation exist among workers, and those in areas of high unemployment seem willing to take on lower-paid jobs with shorter hours. Changes also in education and attitudes mean that young people tend to favour shorter hours at the beginning of their working lives.

Another innovation which could contribute to the revolution in working patterns would be the introduction of sabbatical years, say every ten years, for those in the professions and in teaching, nursing and white-collar occupations, and even, for the few that want them, in blue-collar jobs. Sabbatical years can be seen merely as a way of easing the pressures on those in work by taking advantage of the greater opportunities for leisure offered by the new technology; but government could also introduce schemes to provide grants for companies taking on an extra worker for, say, every ten workers offered a sabbatical year. The public sector again could give a lead to industry by introducing sabbaticals for a wide range of civil servants. Initial financial strains could be met by offering civil servants, on joining, the opportunity of lower pay and no sabbatical, or the option of a docked salary and sabbatical at the end of it. Job loyalty can also be helped by sabbaticals.

The second major innovation that the government should introduce concerns retirement. The job release scheme points the way. The very concept of a retirement age should be done away with. This should be replaced by an age – say as low as 50 – at which retirement on a low level of pension is possible, rising gradually to maximum pension at 60. Receipt of benefit should be made conditional – and this is the key provision – on another worker being taken on by a company. The scheme would thus be self-financing: the saving to the state of an unemployed youngster going off social security more than exceeds the additional burden to the state of financing a partial retirement pension. In addition, the government could underwrite a portion of company pension funds to those workers whose early retirement opens up a new job vacancy. More flexible company pension funds would allow workers who anticipated early retirement to put a greater proportion of their earnings in pension funds and receive a larger pension upon early retirement.

Equally, workers who want to stay on in their jobs after the ages of 60 and 65 should not, as is so often the case, be forced into retirement. This minority has an important contribution to make – even if the effect would be to postpone vacancies that would otherwise occur. We anticipate, however, that far more people would take advantage of earlier retirement than late retirement. It is reckoned that as many as 750,000 jobs could be created by opening up the possibility of retirement at 50 conditional upon the creation of a new vacancy.

The third major approach to the unemployment problem is at the other end of the job market, by addressing the thorny question of pricing young people into work through apprenticeships or lower initial rates of pay. This goes against the grain of virtually everything the unions have fought for over the past century: yet it chimes in with a society in which the employed industrial worker should model his or her remuneration pattern on the employed professional. In virtually every professional and clerical job rates of pay rise from initially low levels with job experience, and industry should be no exception. A radically different approach to training, apprenticeships and future employment is called for.

Specifically, we favour the move to link the level of social security payable to a 16 to 18 year-old with either (a) vocational or industrial training or (b) community work. It is right that a condition of payment of the full social security benefit be that a young person opt for one of these. This would require that government and industry set up, in partnership, a massive vocational and industrial training scheme as an extension of the existing youth training schemes: some 75 per cent of the funds for this would come from government. Supervision of the scheme would be undertaken by an employer, who would be obliged to offer, in exchange for greater funding, the prospect of a job – at least half-time – for those who satisfy certain minimum training criteria at the end of two years.

This should blunt present criticism of the YTS, that it leaves many youngsters at a loose end after two years, and that many firms take advantage of it in taking youngsters on, knowing that they can be dropped at the end of the period. It would make training more realistically geared to the needs of the market. The public sector can give a lead by offering substantial training opportunities on a job-split basis to school-leavers.

The community work scheme would be the second pincer in the offensive to cope with the problem of unemployed school-leavers. We reject the notion of conscription, which exists in many other countries, because of the conscientious objection to it of many young people, and because of the effect it would have in lowering the standards and morale of the professional armed forces. All the same, there is no reason why otherwise unemployed young people should not be obliged to devote, say, a year to serving their country in the community; every opinion poll on the subject suggests that an over-whelming majority of young people would jump at the chance. It would serve to introduce them to the disciplines of work, and would give them a sense of responsibility and of community. Community schemes, which could be under the supervision of local authorities, could consist of caring for the elderly and sick, the environmental improvement of run-down inner-city areas and derelict land, assisting teachers and nurses, etc.

Those who did not want to take on a job, enter a training scheme or enter community service (every school-leaver would have an absolute right to be taken on by the local authority, unless he/she was non-co-operative or disruptive) would be granted a level of social security so low as effectively to force them to remain at home. After the age of 18, however, they would be eligible for the normal rate of social security. This is the current direction of government policy.

In order to assist companies to take on new workers after the age of 18, the government should introduce, in the nationalised industries, apprenticeships with lower rates of initial pay but the guarantee of a job thereafter. This again should be negotiated with the unions as an employment-creating measure, not as an attempt to introduce new lower-paid categories. Union-negotiated apprenticeships, which could create up to one million jobs, should pre-empt employers going ahead and negotiating cheap rates with the increasingly frustrated young unemployed, which is resulting in the creation of a pool of cheap labour.

The problem of unemployment resulting from the new technology also demands a radically different approach to retraining. Here, too, the nationalised industries can give a lead, although in the private sector this is best left to employers, even if the government can help with fiscal incentives. The Japanese example of retraining workers as often as every five or six years to meet changing technologies and working patterns should be studied. In the case of whole industries closing down, the government should study how retraining in new industries can be provided for those affected. The government's £1.4 billion Adult Training Scheme is warmly welcomed.

The old government policy of creating employment should be geared only towards those genuinely on the scrapheap: that is, those aged 45-plus made redundant in declining industries through no fault of their own, who are too old or too set in their attitudes to want retraining, and who are unskilled or semi-skilled. For this group, which number around 700,000 of today's unemployed, a significant and responsible job-creation programme in run-down government services could be initiated at relatively low cost, using materials that are largely British-made, in the field of renewing the housing stock, of school and hospital improvement and in the renovation, for example, of sewers and pipes. There is room, as a consequence of the government's astonishing success in reducing inflation to 5 per cent or under, for a £1 billion programme aimed at providing jobs for the unskilled or semi-skilled unemployed.

We do not delude ourselves that this would be anything more than a job-creating measure intended both to provide some benefit for the community and to tide over to retirement those who cannot efficiently be re-employed. But it is necessary and, at a time of low inflation and controlled public spending, affordable.

We believe that these four measures, taken together, would lay the foundations for the long-term reduction of unemployment and for the adaptation of working practices to the new technology. In the short term, as many as two million new jobs could be created by a change in working attitudes pioneered by government, but followed up in private industry, of the kind described. In the longer term the number of jobs lost through the introduction of new technology would constantly be compensated by the change in working practices. Britain would move towards a society in which the vast majority of those who wanted work could find it, and in which the vast majority that we believe want longer leisure hours could also find it, at not too sharply diminished rates of pay. The alternative – a society in which the employed majority work full time and become increasingly prosperous, while the unemployed but expanding minority do nothing and receive little – is unacceptable.

Britain is considered by many people to be lagging behind other industrial countries. In fact, as the first nation to undergo the industrial revolution, it is experiencing most of the problems of post-industrialisation before the others. The British disease – proneness to strikes, absenteeism, less emphasis on reward – is to a great extent the disease of any industrial society where the achievement of minimum living standards has eased the pressure to get on in life, while those at the top have sufficient consciousness of the plight of their fellow countrymen to seek to distribute some of the benefits more fairly. The introduction of the new technology in a long-established industrial society with a firm code of industrial rights and wrongs and established rights, demarcations and procedures was bound to cause trouble. Yet the opportunity it provides – for a society with both greater prosperity and greater leisure time – should be grasped eagerly. Initially less advanced societies have found it easier to introduce the new technology, because of the lack of established labour practices and because workers are keen to accept any job that saves them from a life of penury. The problems of many such societies lie in the future. Britain has achieved an astonishing degree of social peace by adapting its pattern of work to progress – and can continue to do so, by imaginatively embracing the new drudgery-reducing technology.

3 The Age of Quality
Rt Hon Peter Walker, MP

Whenever this century there has been a substantial rise in unemployment due to a world recession and world financial mis-management, there has developed a pessimism as to whether unemployment would continue as a lasting feature of the economic and social scene. For two centuries improvements in technology and the replacement of manpower with machines has created the fear that unemployment would be ever-increasing as technology developed.

Yet the deep depression of the 1930s did not have the result of creating lasting unemployment. The economic activity of rearmament and, thereafter, the economic activity of rebuilding a devastated world saw to it that for several decades full employment was a feature of our scene.

We have now had twelve years under Labour and Conservative governments of a substantial level of unemployment. The task of any political party that wishes to govern our country must be to see that this trend is reversed and the horrors of unemployment are eradicated. The debilitating impacts of unemployment upon the individual and the family are unacceptable. The divisive nature of a nation split between areas of high unemployment and areas of full employment is unacceptable. The economic absurdity of spending billions of pounds for millions of people to produce nothing is damaging to the objective of providing improving living standards and the eradication of poverty in all of its forms. The Conservative Party never has and never should tolerate accepting an unemployment rate of more than three million.

It is right that the government is spending £3 billion a year on measures to support retraining, to generate employment and to generate enterprise. It is right that the government should, in the community programmes, provide temporary jobs for over 300,000 long-term unemployed. It is right that the government should have extended the Re-start programme to remotivate the long-term unemployed and help them on the route back to work. Lord Young's booklet, *Action for Jobs*, brings to the attention of those who are suffering the ways in which the government can be of help. In the coming months we must make the nation aware of our determination to see that this period of Conservative government will be one of return to the prosperity and the contentment that an economy with full employment enjoys. It will not be enough for us to draw attention to the billions of pounds we are already devoting to this task. It will be important for us to spell out the visions we have for the years that lie ahead.

Labour's programme is not, as it claims, a policy to reduce unemployment over their first few years. Their policy would massively increase unemployment in every one of those years. Their hostility to outward investment would mean that every company with an international presence would base their headquarters overseas. Their policies would also guarantee that firms throughout the United States, in the Pacific Basin countries and throughout the rest of the world that wanted to have a manufacturing and production presence within the European Community would base all of their activities outside anti-free enterprise socialist Britain.

Labour's energy policies would not only massively increase our energy costs but would in themselves put 150,000 more people on the dole queue. Their policy to eradicate the new participation that the people have in former state industries and return these industries to the dead hand of nationalisation would have an adverse effect upon our economic vitality. A depressed oil price has recently created problems and pressures for our currency, but it would be as nothing compared with the full weight of Roy Hattersley and his proposals in depressing the pound. Those who have a deep desire to tackle unemployment must also have a deep desire to defeat the Labour Party.

What therefore are the hopes of full employment that we should be looking to?

Firstly, there is our skill at developing and applying what can be described as 'the multiplier effect' – the manner in which the government can take action to accelerate the flow of investment from the private sector. When I was faced with the knowledge that for the coal industry to succeed in the future it had to eradicate its uneconomic pits, I persuaded the Coal Board that it was essential to create an Enterprise Company in which they could help new businesses and enterprises to come to the coal communities which were adversely affected by pit closures. That Company began its activities in 1984. The government authorised £40 million to be injected by the Enterprise Company into creating new businesses and helping existing businesses to expand. In addition to the £15 million that has so far been invested, a further £85 million has been attracted from other sources, meaning that £100 million of new investment has taken place into coal-mining areas with a government outlay of £15 million. The number of jobs so far created is 10,000. The number of projects assisted is more than 750. On this formula by the time £40 million has been spent more than 25,000 jobs will have been created and more than 2,000 businesses will have been helped. Instead of having a future of declining pits fast approaching exhaustion, the mining communities involved have a range of firms that will give their children a diversity of opportunities in the future.

The 'multiplier effect' was operated by a similar company created by British Steel. It was also applied by Michael Heseltine in his dynamic programme to bring new businesses and activity to Liverpool. And it was applied by Nick Edwards in the redevelopment of derelict areas of the urban environment in South Wales, where he encouraged new businesses to move into new properties on a very considerable scale.

We now have enough experience and knowledge of the multiplier effect to apply it nationally in a far more sophisticated way. We know it is not only important to think of money to encourage other money to come in, but we must think of the use of existing buildings where their former use is no longer required. There is a need to have on hand management and financial advice so that the new businesses of today do not by mistake become the bankrupt businesses of tomorrow. The coal industry has been very successful not just in providing the money for the new jobs but also in providing the workshops, the buildings and a range of managerial services. We need to study carefully the multiplier effect on the inner cities. Many of our worst inner-city areas are in the hands of the militant left. They will never bring economic revival because they have a passionate opposition to free enterprise itself. The militant left have crippled Liverpool, not revived it. The government needs to illustrate how the multiplier effect can bring new hope and revival to our inner cities.

The multiplier effect is the type of partnership between government and free enterprise that we need in a modern world. We know that all of our competitors have created a partnership between government and industry, be it the French, the Germans, the Japanese or the Americans. If they have a working partnership and we do not, they succeed. We must develop successful partnership techniques and apply them to the full.

We became rich by mass producing for the world. We have been

hit in recent years by parts of the world, such as Japan and the Pacific
Basin countries, being far better at mass production than we are. The
new technologies may now enable us to regain some of the mass
production markets, but they are technologies that do not require
people. In Britain we have a historic background of industries originally developed from the craft and cottage industries that we should
revive on an enormous scale in a way which would exploit not just the
local markets but world-wide markets which are available for such
goods. The application of high skill and the guarantee of high quality
is what is needed to revive a number of major industries. To revive
them would mean a great deal of training for those who do not have
the skills. It would mean setting up a world-wide marketing mechanism which in the first place both small and medium-sized firms could
take advantage of. It would mean the possible imposition of a
voluntary quality control whose approval was accepted as a hallmark
of British quality goods.

We should seek a considerable extension of the use of skilled
labour in the production of high-quality furniture, fabrics, porcelain,
glass and, on a smaller scale, in such areas as bookbinding and
cooking. A highly successful bakery in East Anglia has created a
highly successful export trade to Germany. The weight of the Germans of Hamburg is being adversely affected by the taste of the cakes
from Kings Lynn. Britain could become the world leader for quality
in a range of industries every one of which is labour-intensive.

We should therefore examine the possibility of a strategy to
become the world's provider of quality goods. As the new growth
areas of the Pacific Basin, the Soviet Union and parts of the United
States increase their prosperity, as the world tires of the purely
mass-produced, there can be no doubt that the market for high-
quality, individually produced items is going to be a vastly expanding
market. It is also a market where others could not speedily follow.

In looking at labour-intensive activity, we must have a clear idea of
what is necessary with regard to the infrastructure. We have a very
old housing stock that needs restoring and repairing. We have a lot of
nineteenth-century sewers that need replacing. We have communications to some of our docks and ports and around our cities that need
improvement. We have hospitals and schools that need replacing. We
recognise that much of this can only be done when accompanied by
economic growth, but we must also recognise that they are themselves part of economic growth. When we come to the next election,
the public will be anxious to hear what we hope to achieve in

improving the built environment, whether it be in our inner cities for the enhancement of our housing stock and the provision of new homes, or whether it be the improvement of the facilities for health and education.

For eight years we have struggled with the worst world recession this century. For eight years against all of our wishes we have seen the rise of ugly unemployment. Our task is to see that the next eight years witness the success of the free enterprise system, bringing about the eradication of unemployment, the enhancement of our quality of living, and real contentment and happiness for the ordinary family in this country.

4 The Drain from the Regions
Rt Hon Michael Heseltine, MP

There are many causes of Britain's industrial decline and the subsequent unemployment we see today. Recession at a time of rapid technological advance has added a social dimension of unacceptable proportions. The causes of our relative industrial decline are many in number, and the priority apportioned to each is a matter of personal and political preference. Our declining share of world markets, our inability to produce, on time, a quality product people will actually buy at a price that is economic, is a story spanning at least half a century.

Yet there is another contributing aspect that I think is less perceived both in its effects on our manufacturing strength and on the regional injustices that the present level of unemployment reflects.

Three huge incentives from the taxpayer are at work, the effect of which cannot be overestimated in drawing resources away from our trading and manufacturing companies, concentrating the control of the resources of this country in the South of England and the City of London, and thereby feeding the market pull to the South East. This trend has accelerated, if not in part created, provincial decline and urban dereliction.

In 1985/6 the cost to the Inland Revenue of mortgage tax relief was about £4,750 million. This exchequer subsidy to home purchase has created the property-owning democracy which, together with the enfranchisement of the council tenant, is one of the greatest social achievements of post-war Britain. It was designed to give responsibility, a local commitment and a stake in society to people and families. It is at the forefront of our Tory purpose.

What was not perceived was that behind the huge growth of home ownership would come the life assurance companies, riding on the backs of that subsidy by providing the life cover to secure the mortgages. Of course, the building societies grew too, but they were much more concerned with recycling loans within the home-producing industry and thus of much more local relevance. There were distorting factors but not on the scale of the life assurance sector.

The latest annual cash flow into the life assurance market is £6,400 million. This huge annual saving owes its scale in significant measure to the £4.75 million tax relief enjoyed by the home-buying public.

The consequence of such a large taxpayer subsidy works in exactly the opposite way to that intended, when the subsidy itself was designed to spread ownership and enhance local commitment. The life assurance companies – acting in a perfectly rational and prudent way – invest their assets in ways not concerned with job creation in Britain or strengthening our regional manufacturing strength.

They are, to start with, London orientated. Half of the twelve largest are actually based in London. It would be surprising if the investment committees of the rest were far away. But even so, of their total investments only 35 per cent are in the UK company securities; this spans the range of commercial, financial and property companies as well as the manufacturing base. If one compares this with the 24 per cent invested in British government securities and the 16 per cent in property, it is quite apparent that wealth generation is not the prime objective of this form of taxpayer-induced activity.

The second huge subsidy diminishing the strength of Britain's wealth-generating capacity is that given to the pension funds. Another £3,500 million subsidy enables the pension funds to attract the savings of company and worker alike, thereby sucking resources from industry and reinvesting them elsewhere. The pension funds invest just under half their assets across the whole spectrum of British quoted companies. They invest over a quarter in foreign governments and companies and about one-fifth in the British government. In West Germany, companies provide their own pensions. Companies have more to invest, the employees care more about the companies for which they work.

The third direct influence pulling in the same direction is the existence of capital taxation on realisation of gains or inheritance. The family businesses essential to the strength of the provincial economy have for much of post-war history been one-generation companies. In one way or another owners have had to sell; once confronted with the need to sell, the remote publicly-quoted company has been able to offer paper shares with tax-postponed liabilities, whereas the friends, neighbours or members of the local community tend to use cash.

A simple choice: cash and pay tax; shares and delay payment. This is a huge incentive to encourage sales to the publicly quoted company. Its price does not need to be better – the tax system achieves that for them!

The influence of the Treasury, which is pervasive under all governments, has financed much of the concentration of private sector wealth. Britain has been victim of a market force-fed with well over £8,000 million a year. From the viewpoint of manufacturing industry, the smaller provincial companies and the older areas of Britain's declining inheritance, this has proved a disaster. The contrast could not be clearer when one sets this £8 billion against regional aid for industry. For the whole of England today this amounts to only £168 million per annum. The total for Northern Ireland, Scotland, Wales and England is just over £615 million. In three significant ways, therefore – and over a long period – Britain has subsidised a positively anti-industrial environment. In addition, we have encouraged consumption as opposed to investment in our government expenditure profile.

Just take the past decade. Index-linking decisions taken by the Conservatives in the low inflation years of the 1970s for public sector pensions were curiously followed under a Labour government by the Rooker–Wise amendment to index-link tax thresholds for society at large. When one remembers, in addition, that most of the debate about the allocation of public resources took place in the context of no increase in real spending, it is easy to see why the role of governments was so constrained. The amount of new discretionary money available to a Chancellor was tiny in relation to the economy or as a whole. And virtually all the programmes that were based on indexed commitments were consumption programmes.

Before a real debate can begin about the year ahead, most of the options are effectively blocked off, either by index-linking, or by political commitments or by the political pressure groups arguing for this or that programme to be maintained in real terms. The benefit of oil revenues, and too much of our industrial and investment programmes, went down that drain. One can only say that it would have been worse under a Labour government.

To take a politically neutral decade between 1973/4 and 1982/3 (in cost terms):
 – social security spending rose by 61 per cent;
 – lending to nationalised industries – to pay for losses – by 46 per cent;
 – law and order by 45 per cent;
 – health and personal social services by 36 per cent.
Yet spending on transport, support for industry and public housing

all fell over the period. That is not the picture of an economy putting its weight behind a commitment to invest in tomorrow. That is a society consuming as though there was no tomorrow.

For these, and many other, reasons (some of them outside our control) we have a scale of unemployment that is now of permanent political concern. We have been spending £7 billion a year to support more than three million people who have no work – a huge example of conspicuous national consumption. We now have more people unemployed than served in the British army at the height of the Second World War.

An interesting reflection on that statement is that whilst at the end of the 1930s we had high unemployment, by the mid-1940s those same unemployed were seen to be perfectly capable of – and willing to – work. All through the reconstruction of the 1950s and 1960s, if the jobs were there, they took them. Today the jobs are not there. But nor are the 1930s, and we have to look anew at what unemployment means and what we do about it.

The reference to the army a moment ago is useful not just in terms of numbers. It provides a thought about management. To manage a programme which distributes £7 billion a year to more than three million unemployed is a prodigious achievement. Today the system is stretched to breaking point simply to keep the cash flowing.

In order to achieve the entitlement a two or three minute visit to the appropriate office once a fortnight is all that is required. A signature on a form making it clear that the claimant is not working will, in the vast majority of cases, prove sufficient to know that the giro cheque will follow. The recent interview – Restart – proposals by the government are designed to erect a rather higher hurdle; but no-one should underestimate the scale of the task.

In the ideal world we would create job opportunities gradually, providing sufficient openings to enable the employment services and the market to solve the unemployment problem. But the scale of unemployment and the composition of the unemployed in the stress areas are beyond the scope of existing policies.

The twilight world of black economies, the poverty trap, and the virtual absence of incentives to work for many who can see very little increase in cash in the pocket from working are now deeply ingrained in much of our society. Parts of prosperous Britain that twenty years ago seemed able to sustain employment levels of 97 per cent or more today record perhaps figures of only 90 per cent. That says something not only about lost jobs but also about people's willingness to accept jobs, and about the lack of training and educational standards that they need to enable them to cope.

Does all this matter? Yes: first, the cost of this enforced consumption on our national resources is appalling. To pay people to do nothing is about as wasteful a way of spending the nation's wealth as one could conceive. But the second reason is incomparably the more significant and dangerous. What sort of society with what sort of values will emerge from a background in which a high level of unemployment is taken for granted? How long does it take an unemployed teenager to become an unemployable adult? What attitudes to their role in society do such people have? Whose values are they expected to respect? And what will they tell their children? Perhaps worse, what will others tell their children about the unemployed? The hopelessness and despair of the declining inner cities and overspill estates will become prey to the inevitable malignancy of social unrest. So if economically and politically we see such overwhelming grounds for concern, what is to be done?

The market is not going to provide jobs on the scale the present crisis demands. It will find some. That is the first opportunity. There are many things we should do, and are doing, to increase the size of the market and to train people for it. And the number of people seeking work will decline. Yet a decade from now no-one predicts levels of unemployment below two million. Those levels of unemployment will display the same unevenness of geographic and ethnic bias that they do today. That prospect is just as unacceptable as the levels we face today. It will represent imbalances in society just as unfair, tension points just as exploitable.

The next opportunity lies in sharing existing jobs. This is easy to discuss as a theory, but far harder to apply in practice, as one calls for the volunteers to do the sharing. The old advertisement sums it up: communist with own knife and fork anxious to meet communist with own steak and kidney pie. But early retirement, shorter working times and flexible working practices, including mid-life sabbaticals, are bound in any case to feature on the political agenda of tomorrow.

But you remain with the heart of the problem. Are we going to continue to accept that an assumption remains that significant cash is available for no contribution from the recipient other than a five-minute fortnightly visit to the local employment office?

A national answer would be wrong. A more local and regional approach would produce different opportunities. As a first priority, the government should launch a standard-bearing programme to rebuild our provincial inheritance. No attempt to take a firmer approach to the administration of unemployment would be possible in the provincial cities today without first convincing those who live there that a genuine concern exists for their interests. The North–South divide is much misunderstood; but it is deeply engrained. Whilst the attitude prevails, the more carrots and fewer sticks, the greater the sympathy and opportunity.

A national urban renewal corporation acting in England with the determination that the Scottish and Welsh Development Agencies have shown would enable us to achieve in English cities the dramatic success now seen in the East End of London and on the banks of the Mersey. And it would end the increasing perception that a great part of this country matters less and less, or not at all. It would additionally help to relieve the growing congestion and pressure that is spreading so fast in the more prosperous Southern counties.

What today applies to the declining areas of Victorian Britain and inner London does not apply to the prosperous South. Here there is work or the opportunity of work. But the willingness to work, the incentive to work, or the skill to work does not always match the opportunities. And some know only too well how to work, yet pretend they don't.

There are two requirements. First, we need a higher standard of management of this vast human resource. There are too many things that need doing, endless people who could do them, but the two don't meet. We need to reconsider the assumption that if you are expecting to draw unemployment benefits you can do so without giving some of your time in return. The best use of the unemployed's time is almost certainly some form of training or retraining. A key objective is to help people to obtain and keep proper jobs. But that is not possible in a climate where to take them or to stay in them is a matter of only marginal financial concern.

5 Unemployment and the Poverty Trap
Rt Hon Michael Heseltine, MP

Such is the nature of party politics that it is always tempting for political opponents to select the statistics of unemployment in a way most damaging to the government in office. It has been a not inconsiderable achievement of the present government that, while watching unemployment grow significantly in its early years, it has been able to convince the majority of people that the fault cannot be laid at its door.

The levels of unemployment began to move up in the 1970s and broke the then incredible barrier of one million under the last Labour government. In most equivalent countries of the capitalist world the 1980s saw similar patterns of lost jobs measured to within a few percentage points. This government has enjoyed no exception to this trend. But in no sense can it be accused of creating the trend itself.

Few of us who served in government will forget the experience of the first half of this decade. Month after month we witnessed unemployment surpassing previous record levels; the spectre of three million people out of work cast an indelible shadow. Debate continues as to the causes and consequences of that rise in unemployment. The arguments are many and varied. What seems to me indisputable however is that, whether one supports this government, as I do, or not, unemployment has been a creeping blight reflecting deep seated problems on the supply as well as the demand side of our economy. It is a problem that has beset our European neighbours as it has us.

In this country inadequate training, poor work incentives, restrictions on employers, and trade union militancy combined with demand shocks (such as the 'oil crisis') to create a situation of low productivity, rising wage costs, reduced competitiveness in international markets, double-digit inflation, and a growth in joblessness. The perspective of the 1970s is too narrow a backcloth against which to understand the scale of adjustment which industry would one day have to make in its performance. Trends of industrial decline stretching back a half century at least had to be reversed.

The slogan 'Labour isn't Working' in 1979 was undoubtedly the truth. Perhaps it was not the whole truth. It expressed, however, a genuine underlying concern that drastic action was necessary to 'turn around' the British economy. It was not credible that such a drastic adjustment could be achieved by pulling on the old levers of fiscal demand management. The task was perceived as threefold: to restore a discipline to the macro control of the economy, lowering inflation and controlling public expenditure, whilst at the same time stimulating enterprise and initiative with lower direct taxation, deregulation, privatisation and a range of specific incentives. And the power of the unions had to be constrained within the law. It is to the Government's credit that it approached the task with a steadfast determination. The initial pace of the 'supply-side' improvements was slow, but the foundations were laid for the rapid rates of growth experienced in recent years.

New jobs have been created in increasing numbers since 1983 – nearly 1⅓ million in 5 years. The tide of rising unemployment for the moment at least has turned. Today just over 2½ million of our people are unemployed, according to the monthly claimant count, compared with a peak of 3¼ million in January 1986. Each month we now read of near record falls in the numbers registered as out of work. It is true that there is argument about measurement, about changes in the method of calculation and the accuracy of the monthly claimant count. Some of those without work do not claim benefit although fully entitled and others claim while not truly seeking work. At the same time over ¾ million people are participating in training schemes, such as the Youth Training Scheme, or are employed under special employment measures, including the Community Programme. Schemes such as these, whilst of great benefit to those involved, conceal the real fragility of the labour market. None the less most observers would agree that with GDP growth of 4½ per cent during 1987 a large part of the reduction in the figures represents a genuine fall in unemployment.

Ministers are rightly encouraged by this trend which has moved Britain towards the middle of the international league table of unemployment. Opinion polls suggest that unemployment is causing somewhat less concern amongst the electorate. I, too, welcome the trend. Who could not? But my task is to stress the enormity of what still needs to be done. We should not place our heads in the sand, like so many ostriches, preferring to stress the relatively small improvements in order to divert the emphasis from the relatively much larger problem that remains.

At the most optimistic current assessment unemployment will take a long time, perhaps as much as a decade – that means to the beginning of the next century – to fall back to the levels experienced in the late 1970s. And that is the most optimistic assessment. At the other end of the scale some are forecasting that unemployment may even stop falling this year, leaving us in the chilling position of over 2½ million out of work – despite all that has been achieved.

No one believes in the virtues of a competitive capitalist environment more than I do. But for all its considerable merits, the market unaided does not look like it will possibly find jobs at the speed or on the scale the crisis demands. And nor can the falling numbers of young people who will be entering the job market do more than ameliorate the problem. So levels of unemployment, on any foreseeable assumptions, will be as unacceptable in 1998 as they are today, with the sense of injustice and anger much greater and possibly explosive. I simply cannot think it right for politicians to go on making speeches about how things are going to improve if by that they mean we shall gradually move from levels of three million to levels of two million out of work. By doing so they deceive great numbers of people who at present can only expect to remain for many years as deprived, as underprivileged and as firmly enmeshed in an environment of hopelessness as they are today.

I refer to people in those areas where unemployment is most concentrated and, in the absence of intervention, relatively untouched by the benefits of growth: cells of despair, unrecognised by hundreds of thousands of those living in more prosperous localities where work is plentiful. I refer to the older unemployed man in the North and North West, in Scotland and in Wales, facing a life without working in a dream world euphemistically known as 'early retirement'. I refer to the younger person, unemployed for lack of skills, under educated and without opportunity. Especially I refer to that third of all the unemployed classified as long-term unemployed: more than a million people whom many economists consider effectively pushed out of the active labour market: increasingly unattractive to employers and performing no role in moderating the behaviour of those engaged in wage bargaining.

All of this represents an intolerable human waste. It is a waste of taxpayers' money; the measurable cost of benefit paid to the unemployed, according to Government calculations, amounts to around £6 billion per year. It is equally a waste of the moral fibre of our nation to deny men and women the opportunity to contribute fully to the common good. This is the most insidious feature of mass unemployment. It creates an unemployment sub-culture in our society. The existence of such a class or culture is anathema to the spirit of the Tory party as I conceive of it. It disturbs me to witness the emergence of something of the kind in some depressed areas of Britain today. The factors contributing to the vicious circle of decline in underprivileged sectors of society are complex, deep-seated and long-term, but one of the most acute manifestations of this situation is the erosion of opportunity. No factor is more important than a chronic shortage of available jobs. That root is deepest where unemployment is long-established, where children follow parents to the dole queue.

Conservatives who extol the values of the family and its central position in a stable society cannot ignore the harm done to the cohesion of families where unemployment and impoverishment descend to the second and sometimes third generation. Younger generations are growing up in the pincer grip of their own deprivation and other people's relative prosperity. They live just a stone's throw from a more affluent society whose benefits and attractions are brought daily by television before their eyes but yet there is no legitimate way in which they perceive their ability to share in it. They are vulnerable to those who say 'This is a rotten society which offers neither you nor your parents a fair deal. Why cling to its values? Join us on the streets.' By this some mean political protest of the kind familiar to those of us with experience of urban unrest, while others mean the underworld of crime.

Drug peddlars may accept that what they are doing is wrong but see no other way to earn a living. Macho kids take to 'copy-cat' street offences because one more force in society pushes them in that direction. The values and morality of middle-class comfort are more easily comprehensible if you enjoy middle-class prosperity. And so long as we fail to make advances against persistent high unemployment, so long will we encourage a challenge to the values which we hold most dear.

Fortunately, it is precisely from these values that we draw our greatest strength: the individual responsibility, integrity and enterprise that have bought us prosperity in peacetime and victory at times of crisis. But we need to cherish these values and foster them and by this I mean recognising the obligations of society towards the individual as well as emphasising the obligations of individuals toward

society. I believe that only by encouraging such a 'harmony of obligations', which I see as a realisation of the Tory philosophy of partnership, will we commit ourselves to programmes with an ambition
commensurate to the scale of the problem.

What are these obligations? The first point to assert is that the
obligations are two-way. The state does and should maintain a role in
protecting the unemployed, but those without work have an obligation to do all they reasonably can to find work. In stating this I merely
re-state the position of all Labour and Conservative governments
since the Beveridge reforms.

But in three ways the issues are more complicated today than at
any other period of postwar history. The scale of the problem is huge
and the state has only just begun to address the issues of managing it
in a positive sense. Rapidly we are coming to accept that a cash
dispersal system to ameliorate hardship is not enough. As the scale of
unemployment began to rise rapidly in the 1970s the emphasis had to
be on the ability of the system to distribute cash to the growing
numbers of unemployed people. The machinery was overwhelmed.
Very few questions could be asked and the genuine urgency of
dealing with the deteriorating employment prospect took precedence
over addressing any marginal abuse.

But times have changed. Painfully our industrial, commercial and
entrepreneurial world has adjusted and today new jobs are emerging
fast. It is becoming clearer every day, however, that many of these
jobs require skills that are now not widely available. Many of the
unemployed have lived too long with the assumptions of unemployment. The incentives for people to work can be diminished by a
second family income, the black economy and for some the often
only marginal difference in income between working for it or register-
ing for it.

And so alongside the obligation of society to concern itself with the
unemployed there is now a growing determination to look beyond the
simple insurance provision of state benefit. The government is com-
bining this with the right and rigorous management of the taxpayers'
resources by providing a growing number of training, educational and
work experience programmes. The Government is properly not content simply to hand over the money. In today's world that, on its own,
is likely to perpetuate the problem. It avoids the essential question
and the essential responsibility: how do we help the unemployed to
re-enter the world of employment?

I strongly support the evolution of this process. Nothing is more important than the establishment of a framework of management that links advice on job availability and local vacancies with information about the educational, training and community programmes available and with the unemployment benefit system.

But that then leads us to the other set of obligations which it is less fashionable in politics to discuss. What obligations fall upon the unemployed in a society where social provision to underpin their situation is available – partly because they have paid for it themselves – and where the state offers constructive opportunities to train or work?

The classic Beveridge assumption, upheld by all governments, is that the unemployed must be available for work or their benefits will be withheld. No one quarrels with the principles of such a statement. All governments have been committed to it. So the debate has to be about what work, at what rate of pay and with what rights of refusal by the individuals concerned. And the objective should be clearly stated: it is to find work for, and to train for work, people without it. The purpose is to find employment – not finance unemployment. That should be seen as the 'last resort' safety-net or a short-term bridge to protect those deprived of work whilst they move to new openings.

To follow such a path would be to establish in Britain what in Sweden they call the 'employment principle', as opposed to our current benefit or 'cash assistance' system. In Sweden it is considered preferable to ensure the availability of temporary jobs or vocational training for the unemployed rather than passively providing them with cash assistance. Unemployment benefit (at roughly 80 per cent of earnings) is payable for only ten months but by that time the recipient will have been interviewed and offered either a job or a place on a training scheme. If the offer is rejected without good reason the recipient loses benefit (although not that paid to any dependants). This approach is undoubtedly a major reason why only 10 per cent of unemployed Swedes have been out of work for over a year, compared with over 40 per cent here, and why only 3 per cent of the whole Swedish workforce is unemployed.

In some twenty-five states of the USA experiments have been made with an approach which has some similarities in terms of administration but of course operates in a very different cultural environment. Popularised under the title 'Workfare' it varies from minor schemes in some states for a handful of beneficiaries to major programmes in one or two states aimed at the bulk of single-parent families with an unemployed head of household – but not to the unemployed per se. The basic principle is that anyone who has a family to support and is capable of working should not have to rely on unearned state benefits. Those who cannot find employment are required to perform some public service for part of each week, without additional pay, in return for their benefit.

The crucial difference between the major US Workfare schemes and the Swedish approach is that whereas under the former system individuals 'work off' their benefit, the latter provides 'normal' work, paid at the going rate, or training. Both situations, however, conform to the employment principle and it is essentially this that we should now contemplate introducing in some form into Britain as an alternative to the current situation where the unemployed person has only the right to a cash benefit but not a right to work. Our current approach only serves to foster dependency instead of preparing individuals for a life of activity and an enhanced standard of living.

Of course, no one would suggest lifting foreign schemes and transplanting them on to our society. The task is to explore the principles of such schemes and thus establish their relevance to our circumstances. How might a British system of 'workfare' or 'right to work' operate? Could we design in Britain a concept of 'community benefit' linking service and employment to our unemployment benefit provision? The Secretary of State for Employment published the government's White Paper on the new Adult Training Programme early in 1988. This contains much that must be a part of any effective policy designed to tackle long-term unemployment at its roots. I hope that the scale of the programme, providing 300,000 placements for 600,000 individuals each year at a gross cost of around £1.5 billion, will be seen by the Government in its proper context which is that of 2½ million out of work.

The moment the context is quantified the scale of the task to be tackled is clearly set alongside the present government response. It is clear to me that we have to go further. Much work has been carried out significantly by the House of Commons All Party Select Committee on Employment, the Employment Institute and various policy think-tanks. We are short of neither ideas nor estimates for the cost of implementing those ideas.

The Employment Committee calculated that the net annual cost to the taxpayer of providing nearly a million jobs in the Job Guarantee Programme would be £3 billion. The Employment Institute believes that this is an over-calculation and that a more realistic figure would be £1.3 billion. An independent survey by John Burton for the University of Buckingham's Employment Research Centre refines the cost of a workfare system similar to the American precedent down to about £850 million a year.

I would not at this stage recommend so substantial a commitment. And if I had – in some miraculous way – so substantial a sum of money to help the unemployed I would spend much of it improving the conditions that generate market employment in areas of high unemployment. For example, we could double the whole urban programme. I see the right way forward as step-by-step. The first step is to build confidence in the programmes. This is now happening. Only then can the link to continued benefit be created as is shortly to happen with the YTS. The urban areas need not only the quality of work opportunities but the emergence of confidence and trust associated with an enlarged urban programme. The implications of withdrawal of benefit in many deprived parts of the country, where the social conditions and political fragility are as they are today, are still unthinkable. The community benefit approach that I favour calls for local discretion and flexibility. Progress would vary considerably from area to area. What is practical and defensible in most of the South East is not thinkable in today's rundown inner cities.

The first challenge for the government services would be to build up that part of its existing training and educational provision so that those who might realistically gain from training and/or education would draw benefit by participating in such programmes. Included in a loose definition of this opportunity would be job clubs or other positive steps to locate jobs actually available. The essential feature is that the training is of high quality, that the education educates. It is by common consent today that the MSC have made rapid progress in the trading up of the opportunities they have available.

In this way the government fulfils its obligations to the unemployed. It supports them financially and it helps to equip them to find new work through new skills. Can it be seriously argued that, if the training schemes were of wide choice and quality, people should be entitled to opt out without consequence? I believe that what is about to be the rule with YTS would prove as publicly acceptable for a wider range of age groups. Not everyone wants or would benefit from education or training. But again there is now widespread experience in the provision of valuable work experience. The Community Programme is much respected. Ever since I first spoke about the concept of working for benefit the one thing that has hardly been challenged is that there are endless things that need to be done.

To make the proposals acceptable I would be the first to argue that the work available should not be seen as in some way demeaning or insulting. There is no need for that. That does not mean that unemployed labourers can be offered work as managing directors. It does mean that local charities might find themselves able to get full- or part-time help from unemployed white-collar employees. Housing Associations might engage clerical assistance to improve local management of housing co-operatives. It is not difficult to produce an inexhaustible list of things that need to be done.

I accept that a quick response would be that we should pay the market rate to people to do such work. But that has immediate implications for public expenditure and in practice is not going to happen. And if it were to happen it would have the effect of consolidating the new jobs as a permanent feature of the employment profile. This in turn removes the encouragement for people to continue to seek long-term viable jobs for themselves. With the close comparisons that can now exist between levels of pay in the market and levels of benefit for the unemployed it cannot be the purpose of the state to narrow the gaps any further. The incentive must be the incentive to work.

I believe that much of what I am arguing is entirely consistent both with the spirit and the letter of the existing or proposed law in Britain. At present a fully fit claimant with an adequate insurance payment record is eligible for Unemployment Benefit for up to a year plus a means-tested benefit (Income Support) which lasts for an indefinite period. Inadequately insured claimants receive the latter only should their limited means warrant it.

This is not, however, the entire story. The individual must fulfil the requirements of an 'availability for work' test before receiving benefit. Unemployment Benefit can also be denied for up to thirteen (and soon twenty-six) weeks, and means-tested benefit reduced by up to 40 per cent, if the individual has left his last job without good cause or has refused suitable employment or the offer of a training place. In addition claimants face the ultimate sanction of a prosecution under Section 25 of the Supplementary Benefit Act 1976 which makes incumbent upon individuals the duty to maintain themselves and/or their dependants. These are the rules, although in practice they are not always and everywhere strictly applied, either because suitable employment is genuinely scarce, or because it is difficult to establish an individual's honesty and integrity or precise circumstances. In many instances, therefore, officials use their discretion to turn a blind eye to difficult customers, particularly where refusal of benefit might instigate a lengthy appeals procedure. This is understandable. One does not wish to cause hardship amongst the majority of involuntarily unemployed because of the shiftlessness of a minority. It remains true, however, that application of the rules would be much more straightforward were individuals provided with opportunities to work or train.

I believe that we could build upon the government's successful Restart interview for the long-term unemployed so as to assess an individual's circumstances and employment and training needs. This in itself would discourage shirkers or those claiming benefit whilst working in the so-called 'black economy', and would identify those amongst the hard core of the unemployed who are psychologically inadequate and at least temporarily unsuited to the world of work. Clearly these latter individuals will have to be supported and cared for, but it makes more sense to support such people with invalidity payments rather than unemployment benefit. It certainly makes no sense to go on including such people on registers designed to measure those genuinely available for work. They are not – in the normal assumptions of work.

This leaves us with the able-bodied long-term unemployed who claim that they are seeking work but perhaps are over-choosy about what this might involve. I believe that few of our citizens would object to benefit offices exercising their existing right to refuse benefit to people who did not accept a reasonable offer of work or training placement. Such a situation would, of course, be bound to draw political criticism. For politicans on the left anything perceived, however mistakenly, as a change in the benefit system is like cow slaughter to a Hindu; any hint that benefit should be withdrawn in certain circumstances is considered robbery.

We should not let this deter us. The law is the law and the great mass of the people would see the wisdom of our argument. More-over, I would remind my critics that not only does such a system operate in Sweden, which is suggested as the best Western example of practical socialism, but it also closely resembles what Sir William Beveridge, the propagator of our own social security system, clearly had in mind when drafting his famous Report of 1942.

Beveridge proposed, and I quote, that 'unemployment benefit will continue at the same rate without means test as long as unemployment lasts but will normally be subject to a condition of attendance at a work or training centre after a certain period. The normal period of unconditional unemployment benefit will be six months.' From these words it is clear that Beveridge believed, as I do, that after that amount of time on benefit complete idleness, even though supported by an income, would tend to demoralise the individual. With Beveridge, I share the aim of individual freedom supported by an enabling not a dominating or nanny state.

But this is not the main problem. Shirking malingerers represent but a fraction of our unemployed people. For most unemployed individuals the aim is a job – worthwhile, well paid and providing hope for the future. And so unemployment on the present scale requires just the kind of individual and governmental response which Beveridge experienced in the wartime struggle against poverty and want. This is a response that seeks not to coerce but to reinvigorate our workforce, to restore morale and improve skills. It is a response that requires harmony and partnership, a clear recognition of the duty of the state but also that of the individual.

I believe, therefore, that the government should, as a step to a wider debate about these matters, publish a costed projection of the alternatives implied either in continuing to finance large scale unemployment under present arrangements on the one hand or, on the other, with a more ambitious range of state provided work or training more generally available. We need to know what American, Swedish or Danish practices would cost here, what variant of them we might develop here and to what extent there are genuine and practical obstacles to their adoption here.

6 Unemployment, Education and Skills

Jim Lester, MP, Philip Goodhart, MP, and Tony Baldry, MP

Almost everyone agrees that youth unemployment is one of the most serious social problems facing the country. Public opinion polls record this fact with monotonous regularity; and there is no dispute between the main political parties about the importance of the issue.

The figures have certainly been dreadful. In June 1985, the number of registered unemployed under the age of 25 had reached 1,198,000, but during that month Ministers in the House of Commons easily defeated a set-piece Opposition attack on their policies for dealing with youth unemployment.

The government could draw negative comfort from the fact that all the countries in Western Europe are afflicted with a similar problem, and that the percentage of young people under the age of 25 out of work was not worse than in most other countries. In June 1985, the proportion of under 25s in the total number of unemployed was 37.7 per cent; the Common Market average (excluding Greece) was 36.5 per cent.

On a more positive note, Ministers could draw comfort from the success of their Youth Training Scheme, which has proved popular and effective since it became fully operational in September 1983. Ministers claimed that almost two-thirds of those completing a YTS course find a job or go on to some other form of training or education, and Peter Morrison, the then Minister of State at the Department of Employment responsible for the operation of the scheme, could point to a survey of 15,000 trainees which showed that 84 per cent thought that the scheme was worthwhile.

There was general approval of the decision to extend the Youth Training Scheme for a second year, and most large firms seemed ready to co-operate with this change, even though their costs could escalate quite sharply. As the Confederation of British Industry tartly points out, the scheme would be even more successful if some public sector unions had not blocked the development of YTS in central and local government. Peter Morrison argued, with justification, that the government measures to encourage the training of young people 'represent the most wide-ranging programme of reforms that has been produced for many years'.

The government should continue to encourage young people to price themselves into jobs. Ministers reminded their audiences that when the Electrical, Electronic, Telecommunications and Plumbing Trade Union agreed that the weekly pay of apprentices should be reduced from £41.63p to £27.88p, the intake of apprentices trebled.

The government may develop an enthusiasm for work-sharing projects specifically designed for young workers. Work-sharing for the young seems to have been particularly successful in Holland. (In the Dutch public sector and in major Dutch industries, a thirty-two-hour week for thirty-two hours' pay is now the norm for young Dutch workers.) Imaginative training schemes, realistic wage rates, and sensible work-sharing proposals all have their part to play; but the most important cause for optimism is the simple fact that there will soon be fewer young people looking for jobs.

The long-term prospects for softening the impact of youth unemployment are clearly encouraging. In 1985/6, the number of young people reaching the age of 16 was estimated to be 835,000. By 1991, this figure will have dropped to 667,000. In 1985/6, the number of young people joining the labour market was 119,000 larger than the number of people reaching retiring age and leaving work. By 1990/91, the number of people retiring will exceed the number of young people seeking work by 22,000.

All this plainly has substantial implications for the government's training policy. At the moment, the Manpower Services Commission rightly devotes more than 75 per cent of its £2 billion budget to schemes for young people. The shift in the demographic tide means that the MSC should soon be able to devote more resources to adult training. As public concern for youth unemployment changes, there will be an increased demand for schemes which help the long-term adult unemployed, and which are seen to help to reduce those skill shortages which limit growth. So far, TOPS and a cluster of other adult programmes have not acquired the favourable image that YTS inspires. Ministers and members of the MSC are carrying out a running review of the adult training programme.

Ministers should recognise that the success of YTS stems in part from the fact that diverse schemes have been linked together under one recognisable banner. The YTS has gained widespread support among employers, trade unionists and training specialists, because it is a well thought-out, flexible, realistic training programme. It gained acceptance among young people, partly because it was seen to be a big and expensive programme. If Mode A and Mode B had been presented separately, they would not have had the same impact, and might quickly have acquired the sort of negative image that came to blight the Youth Opportunities Programme.

Some of the MSC adult training programmes, such as Open Tech, are stimulating and imaginative. Some are generally worthy, such as many of the TOPS courses. Some bring new ideas to old problems, such as the Management Development Demonstration Projects, and the Department of Education's REPLAN. Considered separately, these schemes, however beneficial, have a limited impact. If all the adult training schemes were presented as part of one major Skill Improvement Programme (SKIP), the public in general, and potential users in particular, might be more aware of all the opportunities that are available.

In every one of the MSC's fifty-five training areas there should be an easily identifiable SKIP Centre where employers and potential trainees can get information about training opportunities. We note that there is doubt about the future of some of the 164 Information Technology Centres (ITECS). A number of these might change their role and their staff and become SKIP Centres.

Ministers must also take a new look at the question of how our higher education system in general, and our universities in particular, are to be fitted into our national training objectives. There still seems to be an awkward gap between the government's genuine enthusiasm for widespread training and its reluctance to provide sufficient funds for an expanding higher education system.

In 1981, the government subjected the universities to more rigorous financial disciplines. This squeeze has produced some of the desired results. Recurrent costs per student have fallen by 2 per cent in real terms in the last four years, and unit costs have fallen even further in grant-aided colleges. Entry standards to universities have also crept upwards: the average entrant to a university degree course must now produce two 'Bs' and a 'C' at A level, rather than two 'Cs' and a 'B'. Meanwhile, the student–staff ratio at universities has shown a small decline, from 9.4 in 1980 to 10.2 in 1984 – a level which is still remarkably high by international standards. In France, Canada and the United States, the university student–staff ratio is more than twenty to one.

There have, however, been real costs to put against these small gains. At a time when Ministers were repeatedly calling for more science graduates to provide a driving force for the new industrial revolution, the number of engineering students at university actually fell by more than 1,000. Meanwhile, the continuing uproar over university finance has frayed the natural links between the Conservative government and many of its natural supporters in the universities. This argument cost the Prime Minister a well-deserved Honorary Degree at Oxford University. It also masked the fact that admission to polytechnics and other branches of public sector higher education had increased by almost 30 per cent since 1979 – a welcome development that has passed almost unnoticed.

The proposals for the universities outlined in the government's Green Paper, 'The Development of Higher Education into the 1990s', seemed designed to perpetuate this row. The number of 18 and 19 year-olds in the population is expected to fall by 33 per cent between 1984 and 1996. If the Government were to cut the number of university places by a similar 33 per cent, the universities would clearly face a crisis of major proportions. In fact, the Government plans to cut student places by half that amount, thus ensuring that the universities will face a perpetual half-crisis.

The government's present expenditure plans will allow for just under 500,000 full-time students in the mid-1990s, and this projection suggests that there will be a continuing squeeze on university funds, by 2 per cent a year in real terms. There are now forty-six publicly funded universities in the United Kingdom (including the Open University). A real squeeze of 2 per cent a year could mean the demise of one small university, such as Exeter, every two years.

If the government relaxed its firm financial grip and provided funding in real terms at the 'pre-squeeze' 1981 level, full-time student levels in the mid-1990s could climb above 525,000. At current prices, this would mean increased government expenditure of £100 million a year. (In 1984/5, public expenditure on higher education was about £3,400 million.) The government should find this extra money. Ministers have rejected the idea of introducing student loans to help defray the long-term costs of higher education, but it should not be difficult to devise a scheme which would require those graduates who have clearly derived substantial financial benefit from their degrees to pay back some of the public money that had subsidised their university education.

If the extra money is provided, will the students enrol in those scientific courses which the government is particularly anxious to see expanded? If the number of scientific students is going to expand, more schoolchildren will have to take their A levels in mathematics and scientific subjects. Perhaps the only beneficial by-product of the long-drawn-out dispute over teachers' pay has been the general realisation that there is a shortage of good specialist teachers of mathematics and science in the school system. The proposed restructuring of teachers' pay scales should provide the government with an opportunity to give a special financial boost to these specialist teachers.

Even if there is a substantial expansion in higher education, however, most young people will not have the qualifications to participate. The government should now modify one of its own rules which positively discourages less-qualified young people to continue with their education. At present, those on supplementary benefit are only allowed to study for twenty-one hours per week. After that, they lose their entitlement to benefit. This is sensibly designed to stop students who are already enrolled in the higher education system from becoming enmeshed in the supplementary benefit labyrinth. There is, however, a considerable difference between a student who has clearly succeeded educationally, and someone in his twenties on the dole, with few, if any, educational attainments.

The long-term unemployed under 30 (in this context, those who have been without a job for six months) should be eligible to attend approved courses at local technical colleges or colleges of further education, etc., without losing their entitlement to benefit. This would enable the younger long-term unemployed to make much greater use of what is often their most easily accessible training facility – the local technical colleges.

Many technical colleges and colleges of further education already provide courses for the unemployed. These courses improve basic skills such as literacy and numeracy. It clearly makes sense to encourage the long-term unemployed to enrol in such courses, rather than cutting their benefit. The recent report by the Audit Commission has underlined the fact that many non-advanced further education colleges are significantly under-used, and that some of the lecturers do little more than twenty-one hours of work a week themselves. Rather than close these under-used establishments, it would be better to encourage the unemployed to use them as a springboard to employment, through skill acquisition.

But even if an expanded higher education system is brought more closely in tune with the vocational needs of the nation, it is plain that the major role in the government's Skill Improvement Programme must be filled by private industry. In order to encourage private firms to expand their commitment to training, the government can choose one or more of three courses: exhortation, the stick, or the carrot.

There has been no shortage of exhortation recently. Many ministerial speeches rightly take up this theme, and there is even speculation that the next round of changes in the government could produce a *de facto* Ministry of Training which would bombard firms with letters and leaflets drawing attention to examples of corporate training leading to improved corporate profit. As Tom King, the then Secretary of State for Employment, wrote in 1984:

> British companies spent on average only 0.15 per cent of turnover on training last year – so little that it hardly shows up in the balance sheet. That is one seventh of the American figure, and one fourteenth of the best in West Germany.

He might have added that in Japan the Nippon Telegraph and Telephone Company enrolled 240,000 workers out of a total of 312,000 in company training courses last year. The giant electronics firm, Hitachi, estimates that its annual company training budget has reached £70 million, or about two thirds of its advertising expenditure. Tom King went on to say:

> It is ironic that every company report and accounts include the most meticulously accurate calculations of depreciation of building and plant and their replacement cost, yet no such assessment is made of the knowledge and skills of those who will utilise these resources. Would anyone suggest that the latter depreciate any bit less quickly than any fixed assets? It is time that investors and the Stock Exchange took an interest in what provision companies are making to maintain and enhance the level of skills and competence of their workforce.

Of course there are pitfalls here for the government. No sensible government wants to underwrite the training costs of industry, if businesses can be persuaded or cajoled to underwrite these costs themselves. But the government itself encourages 'the most meticulously accurate calculations of depreciation of buildings and plant' through the taxation system. It has been estimated that even after the Budget changes last year the full range of depreciation allowances for building and plant available to industry amounted to more than £9 billion.

The 'carrot' part of the Skill Improvement Programme could consist of diverting a portion of this government's support for building and plant into direct encouragement for training.

Over the last few years the government has cut the taxation of jobs by reducing national insurance contributions, and it has also cut the subsidy for the introduction of labour-saving machinery. It should now devise changes in the taxation system which positively encourage training.

The final report of the Information Technology Skill Shortage Committee, headed by John Butcher, implies that one method might be assistance for training consortia, run by groups of companies. In America, the Reagan administration has provided substantial pump-priming funds for this sort of project.

The government's training 'stick' might consist of a national training tax. This has been mentioned in Manpower Services Commission discussion documents, but runs counter to the government's past policy of phasing out industry-wide training boards and training levies. A national training tax would clearly catch the attention of management, but it would add to costs and be unpopular. A judicious mixture of exhortation and financial 'carrot' is more likely to engage the willing co-operation of business in a field where enthusiasm is important.

But even if the level of company training increases five-fold in the next five years, there will still be many people who will not have easy access to any training. For some of those who find themselves in this position, the government's proposed experimental training loan scheme should be of help. Ministers have earmarked up to £10 million for a pilot scheme which will allow some 10,000 training loans to be made. The potential trainees would be expected to put up about 20 per cent of the required finance themselves. The banks involved in the scheme would put up the remaining 80 per cent of the loan, with half of the bank's share being provided by the government. As the Department of Employment notes,

If it is successful, the advantages for all parties would be considerable. More people will be trained in courses of immediate vocational relevance at very low net cost to the public.

There will be wider economic benefit from allowing market forces to affect the pattern of such training. The provision of new courses will be encouraged. The lending institutions for their part will be able to develop a new area of business and use it to attract new customers ... The scheme is intended to cover any kind of course which will improve the applicants' earning capacity more than sufficiently to repay the loan with interest.

A recent MORI poll suggested that rather fewer than one in ten of the working population would be prepared to take out a bank loan for adult training. On the other hand, 62 per cent of those questioned were interested in trying some sort of training, and half of those willing to train said that they were prepared to pay at least some of the costs.

Some pessimists might interpret the poll as an indication that the British workforce is less likely to respond to training opportunities than its counterpart in other countries, but if the proportion of people who were actively interested in training loans turned out to be even less than 5 per cent, that number would still clearly absorb all the public money that is likely to be available for training loans in the near future. The government should press ahead with its pilot scheme. The potential demand for a scheme of this sort could be very large, and it could play a major role in the government's Skill Improvement Programme.

For those without the necessary resources of money, aptitude or ambition to be able to participate in a training loan scheme, a new scheme of training vouchers could have substantial attractions. For some 200 years, educational philosophers from Thomas Paine, Adam Smith and John Stuart Mill to Sir Keith Joseph have been attracted by the idea of education vouchers. As outlined by Milton Friedman,

> Governments could require a minimum level of schooling, financed by giving parents vouchers redeemable for a specified maximum sum per child per year if spent on 'approved' educational services. Parents would then be free to spend this sum and any additional sum they themselves provided on purchasing educational services from an 'approved' institution of their own choice. The educational services could be rendered by private enterprises operated for profit, or by non-profit institutions. The role of the government would be limited to ensuring that the schools met certain minimum standards, such as the inclusion of a minimum common content in their programmes, much as it now inspects restaurants to ensure that they maintain minimum sanitary standards.

The proponents of the only serious school voucher experiment, which was held in Alum Rock, California, argued that school vouchers would 'increase the schools' responsiveness to their constituents, encourage diversity of educational programs, increase parental satisfaction, and ultimately improve the quality of education.'

All those seriously attracted to the idea of school vouchers – including Sir Keith Joseph – have been defeated by the practical difficulties of grafting an education voucher system on to a system of compulsory schooling where there are a limited number of school places. These difficulties, which have torpedoed all prospects of success for education vouchers, do not apply to training vouchers. The introduction of training vouchers could lead to a very large increase in the whole training establishment, both public and private.

As with training loans, the government should run a well-monitored pilot training voucher scheme. The priority group for receiving training vouchers should be the long-term unemployed under 30, who are not able to participate in YTS. A training voucher at a value of up to £1,000 could be redeemable at a recognised training centre in the public or private sector. As with loans, the individual would be expected to discover what attainable skills were most needed in the local market-place, and decide which skills would give him or her the best chance of employment. A training voucher would be non-assignable. Employers who urgently needed people with particular skills might be encouraged to increase the value of the training voucher. The benefit entitlement of the long-term unemployed should not be affected by receipt of a training voucher. It would be valueless to them in cash terms, and they would continue to draw their normal benefit while they were undergoing their training.

There will certainly be teething problems. Next year, ten or twelve of the MSC's training areas could be given a million pounds to spend on pilot schemes. If these schemes prove to be both attractive and effective, then a training voucher scheme could be extended to all the MSC training areas. A combined system of training loans and training vouchers could make the whole training process more sensitive to the needs of the market-place. It could help to solve the national problem of skill shortages; and it could give fresh hope and encouragement to tens of thousands of the unemployed.

Under the leadership of David Young, and then Bryan Nicholson, with ministerial guidance from Jim Prior, Norman Tebbit and Tom King, the MSC has an enviable record in the field of youth training. The shifting demographic tide should allow the government to adopt an adult Skill Improvement Programme which will match the success of the Youth Training Scheme.

7 Jobs: the Need for Immediate Relief

Jim Lester, MP and David Grayson

There is a dichotomy: parts of the economy are booming, while unemployment remains high. This is not a conventional recession – a simple down-turn in the normal economic cycle. Powerful new industries like computers, electronics, genetics and aerospace are springing up at the same rate as the old smoke-stack industries like steel and textiles decline. To quote Alvin Toffler, the American futurologist:

> What's happening is not a recession as such, but a re-structuring of the entire techno-economic base of society. It's like an earthquake that throws up a new terrain. Unless that's understood, no tinkering with interest rates, taxes, wage and price policies, or trade relations, can save us. (Alvin Toffler, *Previews and Premises*, 1984)

We are living through a jobs revolution as momentous as the agrarian or industrial revolutions. Alvin Toffler describes the changes being brought about by telecommunications and other new technologies as the 'third wave'. Francis Pym has argued that 'We are at the dawn of an age so momentous that it will force us completely to revise our attitudes.' Ministers have to be more explicit in explaining the radical changes that we are living through and the dichotomy of a growing economy alongside still growing unemployment.

Francis Pym, in *The Politics of Consent*, explains this in terms of ten average British workers:

> If we look back to 1968 – a time of virtually full employment – those ten people would probably all have been in work. They would each have been working on average 45.8 hours a week, for 48.5 weeks a year and with the probability of doing so for 50 years from 15 to 65. Between them, those 10 people would have contributed 1.1 million working hours in their lives.

> Move forward fifteen years to 1983. At least one of those ten people would have been unemployed because employment had fallen

by nearly 10 per cent. That leaves nine full-time jobs. On average, those nine would have worked shorter hours: about 42 hours a week, for 47.5 weeks a year for fifty years. Between them, the ten would have contributed 900,000 working hours versus 1.1 million in 1968. Yet by 1983 they were producing 22 per cent more than they did in 1968, so the hours required to generate the same output have been reduced by 33 per cent – one third in fifteen years. And that is before we have experienced anything like the full impact of new technology!

Move forward another fifteen years to 1998. Is it far fetched to assume that the same output will be achieved in half the number of hours that it took in 1983? If we apply our current assumptions about employment – 42 hours a week, for 47.5 weeks a year for fifty years – by the end of the century we will need only four or five of those ten people to do the work of all of them.

This is based on zero growth. In the past, we would generally expect employment to rise in line with growth. If a company's output rose by 10 per cent, its manpower requirements would rise by nearly the same amount. But in companies that are technology-intensive – and even many that are not – manpower requirements will be less directly related to variations in output in the future. On OECD forecasts, it is not unreasonable to assume that an average output of 10 per cent will require an average labour increase of only 4 per cent. Economic growth will not fill the jobs gap.

In 1965 British manufacturing industry employed 8.4 million people – 36.6 per cent of all those employed. By the end of 1983, the number of manufacturing jobs had fallen to 5.5 million, or 26.4 per cent of the total workforce. If new technology has the effect on manufacturing suggested, there will be a further 2.5 million jobs lost in manufacturing by 1998. Thus, only about 12 per cent of British workers will be in manufacturing by the end of the century. Even this is likely to be higher than in the United States, where the Harvard Business School expects only 5 per cent of the American workforce to be in manufacturing in the late 1990s.

We are not suggesting by this analysis of job losses that there is a fixed number of jobs. Between 1881 and 1981, the British labour force increased from 12.25 to 26.25 million, but, of course, unemployment did not grow by fourteen million. We can expect considerable growth in the service sector. If we assume an optimistic 3 per cent annual growth rate to 1998, there will be a 27 per cent rise in service sector employment – an extra 3.5 million jobs by the turn of the century; although it should be noted that some economists argue that there is nothing to indicate that services are immune from or less affected by the phenomenon of productivity growth.

The authoritative Institute for Employment Research at Warwick University published projections in August 1985 of employment through to 1990. It expects a modest rise in the net jobs total between 1984 and 1990 of 140,000. But 900,000 of the new jobs are part-time, with 80 per cent going to women. A 1.1 million fall in full-time jobs is forecast and unemployment in 1990 is expected to total 3.3 million. As the NEDC director John Cassels has warned: 'It remains difficult to find any solid grounds for expecting the general level of unemployment to be much lower at the end of the decade than it is today.'

Thus even on optimistic assumptions, economic growth and market forces alone will not cure unemployment. At the very least, we are currently facing unemployment of over 3 million to the end of the century. Some pessimists expect much worse unless we take further decisive action now. Certainly, full employment as previously defined is a thing of the past but, dismal as these projections are, they must not be mis-read to substantiate the third unemployment myth. We must not become resigned or anaesthetised to increasing unemployment.

In 1984 CBI Special Programmes United reported that they did not know a single major city in Britain where the civic and business leaders expected the current unemployment position to improve significantly in their own areas. Privately, many politicians and industrialists talk of high long-term unemployment. Yet to date government spokesmen have generally suggested that Britain will return to full employment. For example, David Young, on his original appointment to the Cabinet, declared: 'We must and can go back to an era of full employment. It would be a sorry outlook if we thought otherwise.'

The future is likely to be more complex than either of these positions. First, the recent US experience shows that rapid job generation is possible. Between 1974 and 1984 OECD statistics indicate that the USA had generated 14.5 million net new jobs. In the same period the European Community excluding Greece, with a larger population and a larger economy than the USA, created no net new jobs.

Certainly it is true that many of the new American jobs are low-skill, entry-level jobs, frequently part-time and with non-unionised employers where pay rates are often low. The American experience must be evaluated and where appropriate applied in

Europe, but without repeating their mistakes. In particular, the Americans have been helped by the existence of a large home market, monetary union, an enterprise culture, a tax and legal environment to facilitate enterprise, and the ready exploitation of new technologies. Equally, we should recognise that America has achieved its job turn-around owing to a massive budget deficit which has sucked money, and ultimately jobs, away from the rest of the world; and that the Reagan government cutbacks have raised the proportion of the US population on or below the poverty line from 11.5 per cent to almost 16 per cent. There is also a major debate in the USA about whether middle-level jobs are being lost in the changes now taking place in the economy. Slavish copying of American experience is therefore to be avoided, but the positive lesson that unemployment rates can be turned round should give the lie to those who have a defeatist approach to unemployment levels in Europe. It should reinforce the pressure for Europe to pull together as a single economy to emulate the USA. In Europe, and especially in Britain, any turnaround is also going to require greater acceptance of changed patterns of work and changes, too, in the concept of having a job and being 'in employment'.

The discussion of the jobs revolution should not be confined to private groups or specialists. Such profound social and economic changes should be the subject of a national debate which the government must lead. Politicians have to explain these trends and guide voters towards solutions. If the government were to do so, they would strike a receptive chord. Many people instinctively recognise that changes are taking place and that they have to happen; but people are frightened because of the uncertainty about where change is leading. We have to recognise that the pace of change is so fast and its scope so pervasive that many are genuinely uncertain. It is the fear of the unknown, and we have to show care and understanding.

Tomorrow's jobs will not be the same as today's, and they will not require the same skills. There is, of course, nothing inherently desirable about working from 16 to 65 for forty hours a week over forty-seven or forty-eight weeks each year. In the future people will change jobs and even industries many times, with intervening periods of unemployment and retraining. Futurologists such as Alvin Toffler and the *Economist*'s Norman Macrae foresee workers working part-time for someone else, part-time on their own businesses, and part-time in the non-cash economy such as DIY; and that proportions of each will vary for different individuals at different stages in their lives. Indeed, this is already happening. Recent labour force surveys suggest that 5.7 per cent of self employed men have a second job, as do 6.7 per cent of male employees who call themselves full time but who work thirty hours or less a week in their main jobs. In the USA it is already reckoned that the average worker will change his entire career, not just his job, at least four times in the course of a working life.

Working life too will change. The work ethic remains a predominant element in Western culture. Having a job is regarded by most people as a fundamental expression of being a whole person. Young people see getting a job as part of a transition to adulthood. But again, given the revolutionary changes, government must prepare people and ensure that legal, tax, social welfare, education and training systems are able to go with the grain of these changes and facilitate rather than hinder them. In particular, the Inland Revenue and the DHSS will need to respond far more positively than they have done to date.

As yet, politicians have not been able to convey a reassuring picture of the future. The belief that if you are not in full-time work you must be on the scrap heap is so deeply rooted that it requires a revolutionary conversion. Yet these changes in the nature of work are no cause for despondency. On the contrary, it is an exciting process. New work arrangements and lifestyles, multi-occupations and part-time activities can fit into a Conservative vision of a more decentralised, open, meritocratic society.

IMPROVING BRITAIN'S INFRASTRUCTURE

There is much worthwhile investment, both public and private, already being made in mines, railways, roads and industry, and more to be made. The deterioration of much of our basic infrastructure goes on apace. Reconstruction costs increase the longer it is delayed. In 1984, a NEDO Report revealed that the maintenance backlog alone would cost £2 billion for hospitals and £5 billion for public housing. The Department of Environment's recent survey of council housing revealed that nearly £20 billion needs to be spent to bring this stock up to standard.

Economies in public as in private investment can be false. Industrial efficiency inevitably suffers. Firms raise productivity only to lose competitiveness through poor communications and infrastructure.

What kind of public housekeeping is it to spend millions on bed and breakfast accommodation for homeless families in preference to employing unemployed building workers to rescue from dereliction empty premises that can house them?

Large sums were found for the Thames Barrage. Other areas also are at great risk to life and property from unsafe sea defences. Our leaking water pipes and crumbling sewers have become legendary, but the waste and the threat to public health they represent are all too real. This public expenditure will directly benefit industry; the argument that it risks crowding out private investment are today minimal. UK expenditure grew 5 per cent last year against a 1.5 per cent increase in production. Risks of public expenditure crowding out private investment no longer apply.

Previous recoveries have been cut short by wage inflation: higher costs and prices, not higher output, have followed attempts to stimulate growth. Much has been learnt during the recession. There is greater realism from managers and workers. The government has had considerable success in restraining pay in the public sector – success which we believe opens a way to a new approach to job creation. We believe it would be possible to negotiate, in advance, reasonable and binding wage agreements for specific essential and job-creating public investment projects, and that this would carry far less risk of wage inflation than would follow from expansion of consumer demand in the private sector, where wage increases are already running well ahead of inflation.

The inflationary dangers would be even less with the type of strategy we are proposing. Extra spending on our cities, and more emphasis on health care and on jobs for the long-term unemployed, imply additional demand just where there is currently greatest slack. With high unemployment in the construction industry, higher demand there would lead to higher output and higher employment, not faster inflation.

There are risks in the stimulation of demand, particularly when the economy starts approaching full employment. But we are a long way from that position at the moment. With a large amount of slack in the economy – capital as well as labour – the chances of an upsurge in inflation generated on the fiscal side are remote. Public sector borrowing on a limited scale and for specific investment purposes is, we believe, entirely justified. No successful company finances investment out of current income for very long.

Most comparable countries have far higher public borrowing than we have – without high inflation and without interest rates at our levels. The pressure on rates from, say, an additional £2 billion of public borrowing would in any case be small in comparison to the pressures from oil prices and the attraction of the dollar. Nor need we worry that borrowing more will increase inflation through its effect on public sector debt. The national debt is already down to half of the level of twenty years ago, and is still set to fall relative to national income. Even keeping it constant at today's levels would allow an increase in public borrowing; and in a recession it is right that the ratio should rise as tax receipts fall and benefit payments to the unemployed increase. The PSBR can be increased without prejudicing confidence in the government's economic strategy.

A £1 billion increase in capital investment is estimated to create 165,000 new jobs, whereas the same amount of money spent on tax cuts would result in the creation of only 30,000 new jobs in the same financial year. Investment in public expenditure for the benefit of the community as a whole – for example on construction and other public works – provides jobs for those who are now unemployed. Those people will tend to spend their new income on goods produced in this country, whereas people already on higher wages will tend to spend more on imports: that is to say, the import content of extra capital investment is much smaller than that resulting from tax cuts. It makes no sense to tell the unemployed that they need the incentive of tax reductions for those in work. These people want jobs. Opinion surveys suggest that even those who stand to gain from tax cuts would prefer that money to be devoted to job creation.

Specifically, therefore, in order of priority, the Chancellor should first increase spending on infrastructure, training and employment initiatives; then concentrate any tax cuts on the lowest paid and on employers' contributions. These should be higher priorities than over-indexation of thresholds or reductions in the standard rate of income tax.

A public investment programme targeted principally, though not exclusively, on construction can be designed to have a relatively low import content and a small negative impact on the balance of payments. A construction programme can be introduced selectively and focused at the micro-level of the economy. This would give a stimulus to the depressed construction industry in those areas which have the greatest social need for new investment. For example, the programme of home improvement grants should be expanded and the extra money directed towards the areas of highest unemployment, where housing needs are often greatest. The eligibility rules should be altered to exclude those who can afford to pay for the work themselves, and the extra money should be spent within a certain time limit.

A similar approach should be taken to the revitalisation of industrial and commercial premises in run-down parts of cities like Manchester and Newcastle. More resources should also be allocated to urban renewal generally. London Docklands offer a good example of what can be achieved with considerable private sector involvement, and we suggest a similar approach in other parts of the country. Why should not private capital be drawn in for other projects, as is proposed for the Channel Tunnel? The US experience with public-private partnerships in urban renewal should be applied here with appropriate tax incentives. It must be recognised, of course, that some public or subsidised investments such as the Kielder Dam have proved white elephants because of the recession, or provided few jobs for a very substantial capital investment, e.g. the Sullom Voe Oil Terminal. It is important, therefore, to look at each public sector investment both on the basis of the numbers of jobs that will be created and on the value to the community.

Innovative American projects to reclaim derelict inner-city housing show the potential of Sweat Equity Investments, where unemployed and poorly housed people can be given vocational skills in building and the wherewithal to improve their own housing.

EUROPEAN CO-OPERATION TO TACKLE UNEMPLOYMENT

Neither the causes nor the solutions to Britain's unemployment are to be found just in Britain. Across the European Community, unemployment in the twelve member states is more than 14 million. In addition to its own national initiatives, the government has to co-operate with our European partners for a co-ordinated economic recovery programme.

The Conservative Members of the European Parliament have already taken the lead in the Parliament in developing the elements of such a recovery programme. The Herman Report, adopted by the Parliament in April 1984, sets out the key elements of such a programme. It is based on a major economic report by two leading European economists, Michael Albert and James Ball. This report points to the squeeze in the market sector in Europe during the 1970s as economic growth and business investment fell behind public and private consumption and sharply reduced Europe's ability to market new products and to create employment. The two economists commend those governments who restrained public and private consumption and emphasise that no single European government, not even the West Germans, could get their economies going alone. However, it is not only possible but vitally necessary to avoid further falls in investment and employment and a progressive deterioration of the welfare sector.

This requires co-ordinated European action along the lines recommended by the Herman Report, namely:

− Consolidation of the internal market: harmonising technical standards, abolishing intra-Community frontier controls, removing restrictions on the provision of services, opening public markets, strengthening controls over state aids.
− Integrated capital markets: abolishing restrictions on capital movements, promoting the creation of a European stock exchange through the fiscal harmonisation of transactions in transferable securities, developing new Community instruments for mobilising private savings.
− Development of a single currency market: improving economic co-operation by moving to the next stage of the EMS, developing the ECU to be a convertible currency backed by its own reserves.
− Targeting Community financial resources to certain priority areas: Community priorities to include energy, transport, communications and research into the new technologies with particular reference to the needs of small and medium-sized enterprises and the old industrial regions.
− Co-ordination of national economic policies: the development of a medium-term financial strategy for Europe with the following goals: cutting back growth of the money supply to a target compatible with inflation rates of less than 4 per cent; reducing of public sector deficits; restructuring public expenditure to favour investment over consumption; holding real wages to a level which will enable companies to increase their profits substantially; reducing the indirect wage component; and increased selectivity in allocating social benefits.
− Adaptation of labour to new working conditions: encouragement at Community level of flexi-time systems so that any reduction in working time does not damage company profitability and competi-

tiveness. Increased support through the Social Fund for vocational training so as to produce a workforce whose skills are geared to the needs of modern industries.
- Internal co-operation: the Community to speak with a single voice in exchanges with the United States and Japan with a view to combating protectionism, securing stable exchange rates and an improved system of managing international debt. Improving relations with the developing countries through increased trade, assistance in making them self-sufficient in food production.

Europe does not need a new Messina Conference as some have suggested. What is needed is the political will and dynamism to take the necessary action.

In the modern world, entrepreneurs are best able to compete where they enjoy a large domestic market: new technology generally performs most profitably where there is a large domestic market giving rise to greater specialisation and longer production runs. The large American home market has been an important factor in the USA's embracing of new technology and in the success of its new enterprises. The advantages of a large home market were in large part why British business supported British membership of the European Community. Yet obstacles remain which break up the internal market of the Community so that there are expensive real barriers to trade between the member states. This puts at risk not only our economic future, but also the future of the Community itself.

The development of the internal market of the Community is the top priority facing the Community. The cost of the present fragmentation and limited scope of the internal market is overwhelming and is directly reflected in lower levels of employment and standards of living throughout the Community than we could otherwise enjoy. We must seek complete internal free trade in the Community.

The Government is already promoting this. It should be given a higher profile and the British presidency of the Community Council of Ministers during 1986 was an opportunity for this. Britain's position would be strengthened in these debates if its general attitude was more European and it was recognised that to secure our goals in the internal market will require reciprocal concessions on other matters which other European allies see as crucial to the development of the Community.

This means that unnecessary bureaucratic obstacles to trade within the Community – frequently a tool of protectionism – must be

removed as a matter of urgency. For example, the Commission have
estimated that the annual cost associated with the delay and bureaucracy of crossing national frontiers within the Community is about £7
billion. In this respect, we warmly welcome Lord Cockfield's proposals for an integrated market by 1992.

If producers are to benefit all those who live in the Community by
taking advantage of the opportunities presented by a large home
market, then in principle goods manufactured in one part of the
Community must be permitted to be sold in the rest of the Community.

The Community's Court has sought to promote the integration of
the internal market, but only a major new initiative in this area will
achieve the necessary results. We therefore urge the creation of a
European Standards Institute so that the convergence of technical
standards can be rapidly implemented in the Community. Wherever
possible, these standards should conform to standards that are
acceptable in the rest of the world, so that goods manufactured for
the Community market can also be exported. Top priority should go
to harmonising those requirements which are most readily used as
protectionist devices.

One of the most unsatisfactory elements of the present state of
development of the internal market is that there is still not a genuine
internal market in the service sector. A programme to achieve a
genuine single Community market in services, e.g. freedom to provide banking, insurance and other services throughout the Community, is needed.

The economy of the Community will achieve its full potential only
when there is close integration of the financial markets, including
venture capital markets in the Community. A Community-scale
financial market will help market forces to direct European capital
and savings towards profitable and productive activities. The development of a Community venture capital market will help promising
small and medium-sized enterprises throughout the Community.

State aids to industry, direct or otherwise, can often be protectionist measures which break up the internal market. State aids have also
been used to disguise the damaging effects of inefficient industries
rather than to assist necessary restructuring. State aids to industries
capable of managing without support from public funds are as
protectionist and ill-conceived as any other measures which fragment
and distort the internal market of the Community. The Commission
should ensure that public procurement contracts are in practice put

up for tender and then accepted on a genuinely competitive European basis. More generally, the Community should pursue the spirit and the letter of the GATT initiative, which seeks a greater element of competition in public procurement policies.

The telecommunications industry provides one example of changes needed. The Commission put forward proposals which would open the purchase of 10 per cent of telecommunications equipment to full competition within the Community and which would promote convergence of standards in telecommunications equipment throughout the Community. Whereas in the USA business has been able to anticipate a vast expansion of telecommunications in its home market in this decade, in the Community there has not yet been agreement on the Commission's proposal, even though over the Community as a whole investment in telecommunications in the next decade may amount to more than £100 billion.

Fragmentation of the European telecommunications industry is merely one example of the way in which European producers are unable to realise the benefits of being as competitive as possible. Most cellular radios, for example, are imported from Japan. Yet new jobs and higher standards of living can be achieved if the internal market is developed.

TOWARDS A NATIONAL INDUSTRIAL POLICY

A more aggressive competition policy does not reduce the need for government to have a clear industrial strategy. Since 1979, this government has actually spent more on BL, BSC and the other nationalised industries than any previous government. It appears embarrassed to have done so. Worse, it has allowed the impression to gain ground that it is not particularly concerned by the problems of British manufacturing industry. Britain's manufacturing trade has gone from a surplus of £2.5 billion in 1982 to a deficit of £3.8 billion in 1984. It is likely to deteriorate further and shows no sign of reversing.

In the decade to 1984 British manufacturing output fell by 4.3 per cent, while Japan's rose by 61 per cent; America's by 42 per cent; West Germany's by 16 per cent and Italy's by 22 per cent. Uniquely among the main industrial nations, Britain has not regained its pre-recession level of manufacturing output. Between 1978 and last year, the number of manufacturing jobs in Britain fell by over 1.7 million, import penetration rose by one-fifth, and as much as one-sixth of manufacturing capacity was scrapped and not replaced. In the past twenty years, the UK's share of world trade has been halved.

The future looks even more disturbing. Britain is an inventive nation – according to the Japanese Ministry of Trade and Industry we have produced 55 per cent of new successful inventions since the war – but continues to find itself at a competitive and technological disadvantage 'invented in Britain, developed in America, made in Japan' is an all too frequent occurrence.

Why? Research and development is underfunded. The OECD and various academic and official organisations in Britain have produced a number of well-researched studies over the past few years on the state of R & D in the UK, Europe, Japan and the USA. According to an OECD study published in February 1983, British companies spent only half as much per employee on R & D as their major European competitors in the second half of the 1970s: we were, in fact, second from the bottom of the OECD league for R & D spending relative to output.

In consequence, many British scientists, researchers and industrialists are very concerned that intensifying competition, lack of investment and continuing skill shortages are leading to a widening of the technological gap between Britain and its international competitors. Despite the (ongoing) improvements in the field of training, there is now a real fear that Britain will enter the 1990s with many sectors of her industry based on obsolescent technology and with a largely unskilled workforce, just at the time when declining revenues from North Sea oil expose our industry to the full force of international competition.

Services will become a more important source of jobs and wealth, but services alone will not fill the gap. As the Association of British Chambers of Commerce has argued:
– most service-sector output is not exportable;
– to compensate for the loss of foreign earnings of a 1 per cent fall in manufacturing exports, service exports would need to increase by almost 3 per cent;
– there is a correlation between manufactured and service exports – one-fifth of the service sector's output depends on custom from the manufacturing sector – so a fall in one is unlikely to be matched by a rise in the other;
– foreign competition in financial and other services is increasing. Whereas Britain's invisible trade as a percentage of world invisible

trade fell from 2.3 per cent in 1979–80 to 2 per cent in 1982, America's rose from 4.7 per cent to 5.1 per cent, Japan's from 1.2 per cent to 1.4 per cent; and France's, in 1980–2, from 2.3 per cent to 2.8 per cent.

A recent Bank of England quarterly survey confirms that, since 1981, the trade surplus on invisibles has fallen. And while exports of services (excluding earnings from sea transport) have grown faster than exports of manufactures, imports of services have also grown faster than imports of manufactures. 'Trade in services', the Bank warns, 'is subject to the same influences as trade in goods', and there is no reason to think that a surplus on services is an 'intrinsic feature of the UK economy'.

Yet the Chancellor states: 'I cannot agree that there is any special cause for concern in a deficit in trade in manufactures.' Reluctantly we share the conclusion of the Association of British Chambers of Commerce that 'the government has very little idea of where job and wealth creating production might arise.' Government needs a more coherent and active industrial strategy which recognises that:

– Even though the proportion of total jobs in manufacturing will continue to decline, Britain still needs a healthy, manufacturing centre.

– The dream of a high-tech nirvana cannot be allowed to permit our manufacturing base to shrink not just in relative but in absolute terms. Hi-tech is important, but balanced economic development is essential. There is no British sun-belt to run to.

– Profits can still be made in mature industries provided that there is investment and precise market segmentation.

– Government can contribute to the health of industry both through creating the right economic climate and through direct catalytic action.

– Britain must become competitive internationally.

ELEMENTS OF INDUSTRIAL STRATEGY

1. Cut through the restrictive bureaucracy: for example, COCOM rules forbid the sale of home computers to the Soviet Union – so Japan is about to clean up the market!
2. Encourage innovation through academic entrepreneurs – entrepreneurship within large firms and by new businesses.
3. Stimulate greater R & D through government funds directed to

specific product areas. The government is to be congratulated that 29 per cent of the DTI budget now goes on R & D and technical assistance in growth sectors, compared with 6 per cent in 1979. This percentage needs to be further increased: for example, why not an experiment matching public and private funds to exploit some of Britain's more interesting inventions.

4. Back EC Commission plans to double EC spending on R & D for industry in co-operative, European, public-private partnerships.
5. Develop management training; encourage younger company directors and more non-executive directors.
6. Encourage venture capital funds for new technology products by offering government loans on easy terms to lever out private funds.
7. Learn from Japanese-style trading houses around the world which are able to collect valuable market data and business intelligence and ensure this is made available to and used by British firms.
8. Apply the lessons of successful market segmentation and differentiation of products by, for example, German firms in mature industries and Japanese firms' 'laser approach' to new markets.
9. At least match the political support given to foreign firms bidding for international contracts. British failures to gain the Bosphorus Bridge contract, and the Thailand bus contract fiasco for British Leyland, demonstrate the need for a more positive trade and industrial policy.
10. Play to our strengths – manufacturing skills, energy independence, financial services, small firms.

8 Rediscovering Europe
Robert Harvey

For more than a decade, the progress of the greatest concentration of democracy, economic prosperity and culture in the world has been blighted by the mutterings of a group of sincere nationalists in Britain concerned with belittling the achievement. There can be few experiences more depressing than sitting through the debates on the European Community in the House of Commons and listening to the nitpickers at work. They ignore the single most important movement in world history since the division of Europe at Potsdam. It is necessary to restate that achievement.

After the Second World War, the European continent had decisively lost its primacy and importance in the world; two superpowers, the United States and the Soviet Union, dwarfed the nation states of Europe, which had been impoverished by war. This process accelerated as Europe's dependencies gained their freedom. Europe's fate could have been to resign itself to second-rank status. Instead West Germany and France, so recently locked in mortal combat, joined together in the Coal and Steel Community which begat the Treaty of Rome. Britain, Europe's third big nation, joined after years of trying to assert its influence as a bridge between Western Europe and America, and as the hub of a valuable, but increasingly disparate Commonwealth.

The argument for Britain's joining remains as strong today as then. In a world dominated by the superpowers, it had become meaningless to claim there was a thing as British national sovereignty in respect of many decisions affecting Britain's interests, in particular over the major issues of world peace. Britain has an important but limited role in Nato and is appreciated there by the United States for helping to defuse the anti-American outbursts of the Continental Europeans. But in any major conflict of interest between the United States and Britain within Nato, or between the United States and other European allies, it is the American view which, with modifications, will prevail. In the foremost issue of arms control, the United States and the Soviet Union talk directly, with little reference to their European allies. If an escalation of the arms race or arms control is judged desirable, it will be taken in the interests of the superpowers first, their allies second. The same applies to superpower interests in the Third World: as the European empires have declined, under pressure from the Soviet Union and the United States, both have moved into the vacuum, the Soviet Union using the appeal of its revolutionary rhetoric and, in some cases, armed force; the United States using its greater economic influence, and much more rarely resorting to force.

The European nation states could not hope to prevail against this. So the first great goal of European unity was that of restoring, through co-operation, Europe's control over the issues that affected its own interests and former empires. Far from political union eroding national sovereignty, the purpose was to regain for the nation states, in the wider community, control over issues that had slipped wholly outside their reach. This objective is still a long way from being attained; but co-operation between European foreign ministers has now become routine, and such increased pull as Europe exerts with both superpowers is a consequence of that co-operation. The passage of the European Final Act, which will strengthen and institutionalise that co-operation, is a further move towards regaining a say in our own destinies. A divided Community would actually have less influence over these matters; it is the pro-Europeans who want to assert sovereignty over issues of arms control and influence in the Third World, and the anti-marketeers who, misguidedly, want to perpetuate our loss of national sovereignty.

The second great goal of the EEC – which unfortunately has become its driving force, a case of the cart going before the horse – is greater economic unity. Again, the point is to regain control over our own destinies. The world economy is increasingly, and at a frightening rate, becoming a global one. Huge multinationals, straddling the boundaries of nations, have become powers stronger than some of the smaller nations. More significant, the creation of global money markets means that huge flows of capital can determine, in the short term, the fate of national economic policies. It is naive to leave these gigantic flows to the working of the international money markets, as they are largely left now. The size of the transfers – in particular during the oil crisis of the 1970s and the debt crisis of the 1980s – can wreak havoc on the smaller economies, and even be outside the control of the larger economies.

Equally it is ridiculous for small nation states to seek to insulate themselves from the outside world through exchange controls and fixed currencies. The only hope of an influence at least equal to that of the United States, the Soviet Union or, to a lesser extent, Japan, is to form a currency block big enough to minimise the damage that short-term speculative flows can inflict. In that respect, European moves towards a common monetary union are desirable, the European Monetary System has been a great success, and Britain's membership is long overdue. Why should the success or not of Britain's economic policies be dictated by forces wholly outside its control – bankers in Zurich or Manhattan? That is a real erosion of national sovereignty. A restoration of national sovereignty would be for Britain, in concert with other equal European nations, to be able to regain control over these international financial movements.

In respect of Europe's internal economy, the process of integration has speeded up trade flows – trade has multiplied 25 times since the creation of the Community. It is also excellent from a strategic point of view, and has allowed Europe to plan its industrial integration with a much greater chance of success than would otherwise have been possible. Had it not been for European co-operation, for example, to regulate which steel-producing areas should be allowed to decline, the collapse of unprofitable ones, in particular in Britain, would have been much faster. It is to be hoped that government co-operation in building up the strategic industries – in particular, high-tech ones – will lead to a faster response to an ever-changing world market. Again, on such issues as protection, the fact of European co-operation has been a great help because, in countering the recent moves of the United States and Japan in this direction, the EEC can threaten retaliation much more effectively than separate nations could. On a level that can be readily perceived by the general public, the removal of customs barriers and financial barriers as proposed in the Cockfield plan will be a massive step forward. When, eventually, Europeans can cross common borders without showing their passports or queuing for hours at customs posts, co-operation will be a reality to many people.

What of the argument that the Community has become a giant, bureaucratic overspender, responsible to no one: a huge layer of bureaucracy interfering with the economies of member states, piling up surpluses on its largest single item of expenditure, the Common Agricultural Policy, but otherwise an expensive and inefficient growth on the body politic?

The argument about Community financing needs to be seen in perspective. The Community budget is tiny in comparison with national spending. In 1985 the EEC spent some 28.4 million ECUS, which corresponds to less than 3 per cent of national central government budgets – or around 20p a day for every Briton (around £65 a year, which is about a third of the average rate bill in Britain, or one-sixteenth of the tax bill). The EEC spends less than 1 per cent of GDP of member states, compared to a total public spending by member states of around 51 per cent of GDP. The growth in Community spending has risen at about the same rate as national spending, so the proportion is not increasing. There is also a major difference between the Community and member states: it must balance the books; it cannot inflate or borrow through deficit spending, or to create resources. So at a time when public sector debt around the Community has risen dramatically, to pay for expensive programmes, the Community has been spending only what is has: public debt increased by sixty times the increase in the size of the Community's budget.

The Community budget should thus be considered in context. The problem of agricultural support is nevertheless a huge one. Spending on intervention has almost doubled in five years, from some 11 million ECUS in 1981 to more than 21 million in 1986. This is well short of the true picture: in practice it appears that the value of agricultural stocks is less than half the book value.

The problem of the CAP is linked to the reason the Community came into existence: an exchange between France and Germany that benefited the farmers of both sides without making allowances for the giant productivity increases in agriculture of the past couple of decades. As this has risen, the size of the surplus has created a very serious problem. But it is not limited to the community: the level of agricultural subsidies in the United States, for example, runs well above those in the EEC. The challenge is to find a way of eliminating the surpluses without further eroding the living standards of those who earn their living from the land. The only clear answer is to provide help for farmers to provide facilities for those who use the land for recreation. Increasingly, the recreational value of land is being understood, and those farmers who lose out agriculturally will have to be compensated adequately.

That, however, is a relatively minor problem seen in the perspective of the EEC's overall spending. The much bigger problem is the question of accountability. In this the EEC has so far failed to live up to the ideals of its founders. Because political progress has been at snail's pace, the institutions have not developed as fast as they should. The most powerful institutions are the Council of Ministers, which reflects the interests of the nation states, and the Commission,

which represents the supranational authority. Up to now, the Council
of Ministers has had a complete veto power over the Commission.
The third institution is, of course, the Parliament. The introduction
of direct elections to the European Parliament has been one of the
most promising recent developments in the Community; but it has
not yet begun to fulfil its promise.

Specifically, the chief purpose of the Parliament should be to
scrutinise, point by point, Community spending. In this it is uniquely
equipped. Its debates on the budget are thorough in a way that is
unknown in the British Parliament. To sit through a session of the
European Parliament on the budget is awesome: it sifts through a
book of literally hundreds of clauses line by line, proposal by
proposal, approving or voting against each. National parliaments,
preoccupied with national issues, are almost incapable of mounting
any scrutiny at all. Specifically, in Britain, Community spending is
discussed in debates which usually consist of diatribes by a small band
of committed anti-marketeers. There is no detailed examination of
spending of any kind. It is wholly absurd for national parliaments,
preoccupied with their own causes, to fulfil the function for which the
European Parliament was specifically created, namely the scrutiny of
expenditure.

This goes to the heart of one of the fundamental contradictions of
the anti-European cause: their contention that the Community is
unrepresentative and bureaucratic. The only way to make it more
representative is, clearly, to strengthen the powers of the directly
elected representatives of the peoples of the member states of the
Community. In Britain, considerable apathy exists, because the
activities of the European Parliament seem remote and ineffectual.
Still, the 40 per cent turnout is not bad compared with local government elections, and the rate of participation is higher in the rest of
Europe. The European Parliament has neither fulfilled the hopes of
its supporters, nor disappointed them. At the moment, Community
spending is proposed by the European Commission (the bureaucrats)
and approved directly by the national governments (through the
Council of Ministers), with only cursory scrutiny by national parliaments. The European Parliament can clearly perform the function of
scrutinising spending more effectively than anyone else.

The European Final Act, in its cautious way, symbolises a giant
step forward in the power of the European Parliament. So sensitive-
minded are EEC opponents that the pro-Europeans have had to
underplay the significance of the change. But the importance of the

act is well, if rather guardedly, described in the House of Commons
Foreign Affairs Committee report on the subject:

> In addition to the extension of majority voting in the Council, the
> Single European Act introduces major changes in the Com-
> munity's legislative procedures which considerably enhance the
> influence, if not the power, of the European Assembly. The new
> procedure is set out in Article 7 of the Single European Act, which
> introduces a new Article 149 in the EEC Treaty. It applies only to a
> minority of categories of Community legislation.
>
> Until now, the European Assembly, despite its powers in rela-
> tion to the dismissal of the Commission and its enhanced powers in
> relation to the Community budget, has had a merely advisory role
> in relation to Community legislation, which has normally been
> made by the Council 'after consulting the Assembly'. Under the
> new procedure the Council will act on a proposal from the Com-
> mission 'in co-operation with the European Parliament'.
>
> The effect of the new co-operation procedure is to allow the
> European Assembly a 'second reading' of Commission proposals
> after a 'common position' has been adopted (by qualified majority)
> in the Council, and to allow the Commission the opportunity to
> adopt the Parliament's amendments before putting a final proposal
> to the Council. At the end of the process, the Council may either
> adopt the Commission's re-examined proposal by a qualified ma-
> jority, or amend the proposal by unanimity. If the Council fails to
> act within three months of the submission of the re-examined
> proposal it is deemed not to have been adopted.
>
> Although the new procedure (which applies only to a minority of
> categories of Community legislation) leaves the final decision-
> taking power with the Council, much will depend on the extent of
> co-operation between the European Assembly and the Commis-
> sion: for the new procedure will require only a qualified majority in
> agreeing to the proposals of the Commission (amended, where the
> Commission agrees, by the Assembly), but unanimity to amend the
> Commission's proposals. The ball is therefore very much in the
> Commission's court: if it can reach agreement with the Assembly,
> its hand in relation to the Council will be greatly strengthened.
> There is, in our view, little doubt that, if the Commission and the
> Assembly are in accord in respect of any particular item of legisla-
> tion, the freedom of the Council to adopt a different position will
> be inevitably circumscribed as a result. We agree with the Euro-

pean Communities Committee of the House of Lords that by this time a 'proposal will have acquired a political momentum which will make a qualified majority in its favour likely and also make the exercise of the veto politically difficult'.

The new legislative procedures established by the Single European Act – including both the extension of qualified majority voting and the new co-operation procedure – will undoubtedly reduce the ability of individual Member States to pursue their national interests to the point of actually obstructing the progress of new legislation regarded as acceptable by the majority. The completion of the internal market is regarded by HM Government as so important a matter of UK national interest that it is clearly prepared to accept the risks inherent in the new legislation procedures. According to the Foreign Secretary, the implication is that all Member States 'have got to be prepared to make mutual concessions'. . . .

Title III of the Single European Act deals with European Political Co-operation (i.e. foreign policy co-ordination between the Member States). We took evidence on this subject from the Foreign Secretary last December and from the Minister of State in early May.

The avowed aim of European Political Co-operation is, according to Article 30(1) of the Treaty, to 'endeavour jointly to formulate and implement a European foreign policy'. Article 30(2) of the Treaty contains undertakings by the Member States to 'inform and consult' each other on foreign policy matters 'of general interest', to take full account of the positions of their partners before taking foreign policy decisions, to avoid actions which 'impair their effectiveness as a cohesive force in international relations or within international organisations', and to develop and define 'common principles and objectives' which will 'constitute a reference point' for national policies.

To these ends, the Treaty provides a framework for regular meetings between Ministers (Article 30(3)(a)) and officials (Article 30(10)(c)) and establishes a small secretariat to assist the Presidency in the administrative aspects of European Political Co-operation. The Treaty enjoins the Member States to endeavour to adopt common positions in international organisations and conferences, including those in which not all Member States are participants (Article 30(7)), and records the willingness of Member States to co-ordinate more closely 'on the political and economic aspects of security' (Article 30(6)). The Commission is to be 'fully associated' with European Political Co-operation, and the European Parliament is to be 'closely associated' with the process.

The Minister of State told us in evidence that, so far as European Political Co-operation was concerned, 'nothing new is actually happening. All we are doing is formalising the way in which we have worked over a long period with a slightly larger secretariat'. The Secretary of State similarly told the House in April that the provisions merely translated 'into a binding international agreement the habits and practices that have grown up on political co-operation in recent years'. But in evidence to us in December the Foreign Secretary, having claimed the political co-operation text as a United Kingdom initiative, went further in saying that 'it is not just a codification thing . . . it is to enhance the legal clarity of it and, hopefully, enhance the attention and respect which is paid to it'.

Although we accept Ministers' assurances that 'there is no provision for sanctions for failing to comply' with Title III of the Treaty, we believe the attention of the House should be drawn to the much greater political commitment to foreign policy co-ordination and co-operation which this part of the Treaty represents. It is one thing for sovereign states to informally agree to consult on foreign policy issues (as has been the case to date): it is quite another thing for those states to give written assurances to each other that, for instance, they will consult before they decide 'their final position' on any particular issue (Article 30(2)(b)).

Where should the Community go now? In our view its progress has been marred by overcaution, the result of trying to placate EEC national interests. In particular, France and Britain have been guilty of holding up the Community's integration. The Germans, who favour closer Community integration, have been slow to prod the other two major countries to move faster. The Mediterranean countries are, by contrast, pressing for further integration: and the EEC's enlargement to include Spain and Portugal will give Italy further support in its attempt to create a genuinely supra-national Community. The Benelux countries have a vested interest in integration; although very active, little notice is taken of their views in wider Community councils. The introduction of formal majority voting will make a difference in speeding up decision-making processes that might otherwise seize up in a Europe of the Twelve. In addition, the extension of the consultation process between EEC prime ministers has had a major effect in securing greater co-ordination on matters of common external concern and foreign policy.

The six major areas for further progress are, in our opinion, largely political. These include a strengthening of the powers of the European Parliament; closer co-ordination on foreign, defence, financial and trade policy; the creation of a responsible, elected European executive; the setting up of a European Development Bank; progress towards a Europe of the Regions – stopping a long way short of continental European dreams of federalism; and progress towards a European defence community.

To take each in turn. We believe that for the community to erase its image as a remote bureaucracy the power of the Parliament must be strengthened. That requires a number of minor things: first, the Parliament must meet in one place, not two, so that longer and more thorough sittings can take place. Second, it should be impossible to sit in both the national and the European Parliament; it confuses responsibilities, as well as making members less effective in the European forum. Thirdly, and in the longer term, the Parliament's right to scrutinise EEC actions should be extended to a power of rejection of detailed budget provisions; it is not enough that Parliament has only the power to approve or throw out the whole budget.

A much more difficult question is whether the EEC should have the right to levy spending of its own; if it acquired that right, a much greater awareness of the importance of the EEC would be created among peoples, and a much greater degree of participation. People might feel inclined to take an interest in Community affairs and to vote positively – or negatively – if they see their Euro-civil servants, responsible to national parliaments, enacting taxation and national spending decisions of their own, not merely at the behest of national parliaments. Similarly, national governments know that the moment they concede to the Parliament the right to levy taxes, they will lose most of their hold, and their blocking power, over the development of the Community. The Commission has long been pressing for greater revenue powers, the national governments resisting. The provision of own resources – a 1 per cent slice of VAT – is a formal recognition of the EEC's formal right to raise its own money, but only within the tight margin allowed it by national governments. A very proper question is whether an institution so large and remote from the lives of ordinary Europeans might feel less responsible in spending its own money, for example, than national governments: a big additional tax burden could be the result. We believe this fear to be misplaced: the EEC could be allowed to levy the full burden of its current spending from its own resources with an extra 1 or 2 per cent levy on VAT, and the control the Parliament could exercise over national spending would be greater than any exercised by national parliaments over monthly, horse-traded EEC spending.

The co-ordination of foreign and defence policy will be moved forward a great deal by the European Final Act; ideally, it will mean that over major issues of foreign policy, even short-term ones, an immediate co-ordination beween the highest levels of foreign ministries is possible and that, where possible, the EEC will lend its support for the foreign policy actions of any of its members. Pooled embassies in some countries might be desirable. This would make the EEC a more effective foreign policy instrument than at present. Within the European framework, the Community should not be afraid to take initiatives in respect of the Soviet Union that are independent of the United States. The Soviet Union shares the European continent, and the European Community has a clearer understanding of its concerns – as well as the dangers it poses – than the Americans have. The United States would be only too pleased if the EEC were prepared to set out its differences fully, instead of complaining behind its back at the excesses of American foreign policy. In the field of monetary policy, we favour making progress towards monetary union, which can only be beneficial to all countries, stimulating trade between them and giving Europe greater financial muscle between the two superpowers. With regard to trade barriers, we strongly support the bulk of the Cockfield proposals.

Thirdly, we believe that an elected European executive has now become a necessity. To a large extent he (or she) would be no more than a pinnacle of the bureaucracy. But he would have an authority in world counsels matching, if not exceeding, those of the leaders of the United States and the Soviet Union. He would capture the popular imagination in a way that the nominated EEC Commission cannot, and would personify Europe abroad. A short term – four years – would be required to ensure that the Community was not dominated too much by a single nation, and a two-term limit should be imposed on presidential terms. A two-tier election would be necessary, to ensure a run-off between leading candidates and prevent every country putting forward its own nominee, which the biggest country would win.

Fourthly, Europe's presence in the Third World would be greatly enhanced by the Creation of a European Development Bank, with as large a capital base as the World Bank and the Inter-American Development Bank, capable also of channelling commercial bank resources to the Third World.

Fifthly, the success of regionalism in Spain and Italy, also of federalism in West Germany, as well as the plethora of small states in the EEC, should make the goal of a Europe of the Regions an attractive one. In Britain and France, too, greater regional representation could be achieved within a Community framework.

Sixthly, there should be progress towards European defence integration; this must always go on in concert with NATO; for the foreseeable future, Europe cannot lessen its reliance on the United States. But both our major ally and the Community would benefit from a policy which would allow our countries to stand on their own feet in defence matters. One immediate step should be the creation of a European intervention force to help policy disputes, when requested to do so by friendly governments, in areas of former colonial influence. This would be welcomed by the United States, which is weary of the burden of so doing around the world.

Britain's future, and its hopes of regaining a real influence in the world, lies wholly within Europe. We believe that European civilisation, in its concern for the individual, for the Community, and in its civilised standards, still leads the world; and we believe that this should be so in fact, as well as in the hearts of the European people. A world in which Europe is the greatest of the superpowers will be a more peaceful and civilised world than that of today.

9 Britain's World Responsibility

Jim Lester

No Conservative policy for full employment can ignore the wider world. Clearly, the impact of changing patterns of world trade has an enormous impact on potential markets. Most recently, Britain has come under fierce competition from low-wage competitors in the developing world, which have challenged traditional manufacturing industries. Higher-tech competition from Japan and from some Pacific Basin countries has also had a major impact on manufacturing. Britain has had to turn increasingly to service industries and to higher quality goods to compete. The potential in the higher quality area is immense.

Yet Britain has always, rightly, resisted protectionist pressures and accepted its international responsibilities. As part of the European Community, it has sought to defend its interests from aggressive market penetration by some new competitors, while pressing for the lowest tariff walls possible. The newly industrialised countries should not be seen as a threat to an ailing economy, but as an opportunity for a healthy one. This calls for two things, in particular:

1. A policy of promoting overseas aid of the right kind, firstly on humanitarian grounds, and secondly because it wins Britain friends abroad and ultimately develops Third World economies, offering further trading opportunities for this country.
2. An initiative by Britain on the subject of Third World debt, extending the Chancellor's existing initiative for the poorest African countries, that lessens the burden of a debt that should never have been allowed by international regulatory institutions to build up in the first place. The Third World debt problem is the single most serious problem affecting the Third World today, and is leading to a decline in the West's market with Third World countries, which have sharply cut back on their imports in order to service the debt.

How can these two objectives be achieved? First, the notion that aid is in any way harmful to the recipient must be decisively rejected. Of course, the wrong sort of aid can be bad. But, as Professor Roland

84

Cassen has attested in his evidence to the House of Commons Foreign Affairs Committee enquiry on bilateral aid, the evidence from his own exhaustive enquiry suggests that the right sort of aid is conducive to economic growth:

> We rather take our academic colleagues to task for having done somewhat unsatisfactory international statistical investigations where they have tried, usually rather crudely, to correlate aid with economic growth. We have found most of the studies which have attempted this in the past to be wanting in varieties of ways. In particular, what is usually done is to lump all aid together in one figure and call it 'aid' and to relate that to various measures of economic performance. Now that is mistaken in our view. Aid has various purposes and it comes in various forms. Some aid is very clearly of an investment nature, particularly that which goes into investment projects, such as infrastructure projects, and has a very long gestation. Other aid is what might be called consumption aid, food aid; there are some kinds of budgetary support which are spent very rapidly, and other kinds of commodity assistance.
>
> Then there is another kind of assistance, which is very large in relation to the total, namely technical assistance. Lumping that into the equation, that is even more of a subject of concern because, for one thing, the periods over which technical co-operation works vary enormously, and, secondly, a large amount of the cost of technical co-operation is not actually transferred to the developing countries. The Swedish timber expert on loan to Tanzania spends only a small fraction of the cost of his co-operation agreement in the country. Quite a lot of it is his personal salary, which may pile up in a bank account in Stockholm and various other costs. So that because a country is receiving £100,000 of technical co-operation, that does not mean it is receiving £100,000.
>
> Therefore, for various reasons, one ought to try to separate these different types of aid and their uses when relating them to economic performance. Then the whole way in which the relation of economic performance is calculated in some of the studies also struck us as unsatisfactory and we were particularly pleased to find one study which looked at relations between aid and savings. There were two fairly well-known studies previously which suggested there was a negative relation, that the more aid a country received the less it tended to save; whereas this study was conducted doing exactly what we had suggested ought to be done, namely to

discriminate between the various types of aid – investment aid and consumption aid, so to speak. This study found that when you broke down aid in this way, the investment use was positively correlated with savings, not negatively, as was previously assumed.

So we thought that the studies which alleged that you could not see any relationship between aid and growth were unsatisfactory and we also felt that in the countries we looked at – at least in some of them – you could see good relations between aid and economic performance, though usually differing in each case. So our general conclusion was that one should look at this on a case-by-case basis.

Second, the experience of the famine in Africa between 1983 and 1984 was that Britain must have a larger budget specifically available for emergencies of this kind. The resources available for developmental assistance have been allowed to be squeezed by the huge expansion, because of the famine in Africa, in resources available to humanitarian relief (mostly emergency food aid and refugee relief). The Overseas Development Administration (ODA) budget was not increased to allow for expenditure on the African crisis. Presumably this was because the government took the view that developmental assistance and humanitarian relief were so closely related that the funds originally allocated for development were being properly spent if they were instead transferred to humanitarian relief. This may occasionally be the case; however, less money should not be made available for development just because in an emergency more is needed for humanitarian assistance. In future the two objectives should be treated separately. The development budget should not be raided to provide funds for humanitarian assistance.

The main purpose of British aid must, of course, be long-term development: the promotion of growth and aid to the poorest. Britain's ODA defined the fundamental purpose of the aid programme as being 'to promote sustainable economies and social progress in developing countries, recognising that aid has a particularly valuable role to play in assisting the poorer countries', and its essential nature as being that of concessional finance in a form which can be 'particularly important for the poorest countries'. For ODA, promoting development in this way did not mean their 'looking purely at the GNP consequences of undertaking some activity; we would be interested in factors which could not be captured by quantified economic analysis'. These wider social objectives include the effect on the poor, on the environment and on the position of women.

However, where the relief of poverty is discussed as an objective, the stress in the ODA's evidence is always on the giving of aid to the poorest countries. The Minister of Overseas Development, Mr Chris Patten, in evidence to the House of Commons Foreign Affairs Committee, rightly claimed credit for Britain's position near the top of the list of the proportions of donors' aid given to the poorest countries. This is, however, a different proposition from the thrust of the 1975 White Paper, which stressed that British aid 'should directly benefit not only the poorest countries but the poorest people in those countries'. Furthermore, there are a number of countries whose per capita GNP does not place them among the poorer countries but which have substantial groups of poor people within their population. The ODA recognised the importance of examining the extent to which the benefits of aid would be distributed within different sectors of the population, but regarded the per capita GNP of a country as the starting point for assessing its degree of poverty and thus its eligibility for aid funds.

But even more than in assessing country allocations, it is in assessing proposed aid projects that the priority to be attached to assisting the poorest people must be emphasised. The thrust of the ODA evidence was that the main purpose of the bilateral country programmes was to promote growth and thus indirectly to improve the lot of the poorest people, rather than directly to attempt to relieve poverty. This is a most important difference: the issue of how far the benefits of economic growth will 'trickle down' to the poor was described by Professor Cassen as one of the 'grand questions' in the aid world. Lord Bauer distinguished sharply between the objectives of helping the poorest and promoting development.

ODA took the view that the two objectives – generation of economic growth and relief of poverty – were not only compatible but closely interrelated, and that ODA priorities and procedures already laid stress on the impact of British aid on development in the rural areas (where most of the poor are). The Minister for Overseas Development has expounded on this point at greater length. While readily accepting that there was a need for direct action on poverty, he laid stress on the need for wider economic growth as a likely prerequisite for any sustained and substantial effect on the position of the poor. It is on this basis that the prominent place currently accorded in British bilateral country programmes to infrastructure projects in the poorest countries (for example in the power and transport sectors) is justified by ODA as a contribution to the needs of the poorest people.

But it is by no means clear that the promotion of general economic growth is a sufficient objective. Professor Cassen reported his study's conclusion that there was a 'variety of experience' on the issue of whether or not economic growth generated real benefits for the poor, and that 'there is no automatic reason for growth to turn into income for poor people and no automatic reason why, even if incomes grow, people's nutrition necessarily improves'. Professor Cassen concluded that there will always be situations and types of poverty which must be tackled directly if aid agencies are to achieve the objective of relieving poverty.

But close scrutiny must also be made of proposals for projects which are not directly aimed at the poor to ensure that they do not cause harm or are not missing opportunities for helping to relieve poverty. This conclusion closely matches that of the Members of the Independent Group on British Aid, who felt that – partly because of the increased weight given to commercial objectives – inadequate attention was being paid to the need for poverty-related projects and that sectoral priorities were being distorted. Of course, it may well be not for macroeconomic reasons that the beneficial effects of aid-inspired economic growth do not reach down to the poor: it will often be because of obstacles in the social or political structure of the country concerned, including corruption (a matter we consider in the next section).

Professor Cassen stressed, however, that the use of aid for a direct attack on poverty should be only one of three developmental purposes of aid. The others were 'to contribute to long-term development, especially infrastructure ... and to try and help the growth process in that way'; and 'to try to assist countries with poor balance of payment prospects' together with policy-based lending. The correct approach to the achievement of the developmental objectives of the aid programme cannot be narrowly defined as the direct relief of poverty, and will be different in different countries and situations. The promotion of general economic growth is therefore an important aid objective. Nevertheless it is important to ensure that too much is not taken for granted in assuming that the poor will benefit from general economic growth or that sustained growth is a prerequisite for an improvement in the lot of the poorest people. It remains essential to examine the impact this growth will have on the poor.

I believe the ODA should continue to give priority in their bilateral country programmes to the needs of the poorest people both through:

(a) continuing the development of techniques for the examination in all project proposals (and also in evaluation studies) of the effects of the project on the poorest people in the country concerned; and

(b) giving greater priority to projects designed to attack poverty directly and to be sustainable by the local population after completion of the aid element.

THE POLICY AND ADMINISTRATIVE ENVIRONMENT IN THE RECIPIENT COUNTRY

ODA policy and objectives in setting the priorities and the overall allocation for each country must, however, be influenced by a further, most important consideration. This is the degree to which the recipient country is able, or can be made able, to use the aid given in a productive way. This can be affected both by the policy environment in the country, i.e. the degree to which the politics of the country and its economic and social policies are conducive to sound development, and by the administrative capacity of the country to implement development projects. Aid given in the past has fallen foul of both of these obstacles. An important consideration is the extent of corruption, or political and administrative practices, which might lead to the diversion of aid away from those it is principally designed to assist.

The policy environment of the recipient country is something to which Lord Bauer, among others, has drawn particular attention, suggesting that the long list of policies inimical to development being pursued by many developing countries made those countries inappropriate recipients of British aid. This is a matter on which many of our witnesses laid great stress. It was, however, suggested that this danger was decreasing to the extent that donors, including the UK, have begun to pay much greater attention to these issues than previously. The ODA stated, for example, that an assessment of a country's ability to absorb aid productively now formed a specific part of their annual review of each recipient's aid needs and of ODA's policy. Such factors also form an important part of the list of criteria against which individual projects are assessed.

Once obstacles to sound development have been identified, there remains the question of how to react to them. One possibility is to devise, in co-operation with the recipient government, aid projects which directly address the problem. In the case of administrative weaknesses in the recipient administration the solution may be for a programme of technical assistance, such as training or the provision of an expert, and indeed ODA described this function as one of the major purposes of the Technical Co-operation programme.

A similar, though less direct, approach is to make aid available on certain conditions, for example on condition that certain economic or social policies are changed in a way which it has been assessed will make the aid provided more effective. This can only come through a process of constructive policy dialogue with the recipient country concerned. This raises the prospect of a charge of interference in the affairs of the country in question (though it has been suggested that this danger was more remote the longer developing countries had been independent and the greater their economic difficulties).

ODA witnesses laid great stress on the importance of policy dialogue and the 'policy based lending' which can result from this. They explained, however, that much of this was done not bilaterally but through wider negotiating fora, in particular the International Monetary Fund (IMF) or the International Bank for Reconstruction and Development (IBRD, also known as the World Bank). Recent cases of UK aid becoming available following a country's agreement to a package of policies in this way have included Ghana, Gambia and Tanzania. In such cases the policies in question typically included measures to free the market mechanism (for example, on prices) and to reduce state controls. We noted that ODA in many of the negotiations between the recipients and the IMF/IBRD appeared to make little attempt to come to an independent assessment of the needs of the country and any changes in policy which might be required. This in part reflected the close liaison which was maintained between the UK and these organisations. Criticism was expressed of this approach by one witness, Professor Mosley, who felt that ODA should always make their own assessment.

There will still be cases, however, where Britain will be asked to provide aid, or may have an existing programme, in circumstances where the prospects of the aid being effective seem doubtful. Where there is an existing programme there is little chance (except perhaps with very small countries) of Britain alone achieving any policy changes through a threat of withdrawal of further aid. As with the question of how far the UK should maintain an aid programme with countries with a poor record on human rights, every effort must be made to use British aid as a 'carrot' rather than a 'stick'. Nevertheless, it cannot be acceptable to allocate limited aid resources to countries which will not use it adequately.

THE DEBT CRISIS

The debt crisis is, or rather ought to be, the greatest moral outrage of our times, far eclipsing – because it is man-made – the African famine, which was largely an act of God, and over which the response of the international community was pretty good. Stated baldly, over the past four years, the indebted poorer part of the world has been transferring substantial resources to the richer.

The squeeze is worst for the poorest of the poor: Africa's debt, officially just over $80 billion, but more likely $150 billion, has increased between 1973 and 1983 by an average of 22 per cent a year – far more than any possible increase in exports, and hence in Africa's ability to pay.

Because of this, World Bank projections suggest that Africans cannot expect to see any improvement in their living standards during the coming decade. In fact the decline may continue.

The straightforward reaction to the news that most of the non-Asian developing countries have, in fact, been undeveloping over the past few years because of their indebtedness is to say 'tough, but they brought it on themselves by borrowing too much'. This, though, is to overlook the extraordinary circumstances in which most of the loans were made. The debt crisis stemmed directly from the 1974–5 and 1979–80 oil price rises, which caused huge burdens to be placed on the balance of payments of the non-oil developing world. Some of the money thus gained by the oil producers went on building up their economies; but much of it went on deposit to commercial banks, who found themselves awash with funds.

Where could the banks lend the money? There were few takers among the developed countries, which sensibly reacted to the oil price hike by squeezing their economies in order to prevent inflation taking off. But the struggling Third World was eager for money, and the banks were eager to lend, irrespective of the risk that lending on a very big scale involved. Many of the loans – and the lenders knew it – were lavished on plainly non-economic projects, or consisted of thinly disguised support for an ailing balance of payments. Some even went straight out of the borrowing countries to buy condominiums in Miami. It didn't matter to the banks, because it was believed that countries could not go bust.

At the very least, the banks – and the governments that failed to regulate this – share equal responsibility for this orgy of lending with the borrowers. The whole thing went sour when Mexico veered towards bankruptcy in 1980–1, and an international rescue had to be staged in August, 1982. Brazil, Argentina, Venezuela and others were quick to follow into the arms of the official receiver, the International Monetary Fund. Repayment periods were extended and interest rates lowered on condition that these countries shook themselves into some kind of economic order. This they did, slashing their import bills and cutting their huge government deficits.

These cuts were made at a drastic cost to the countries concerned. Living standards in Latin American countries as a whole are down by nearly one-third. In countries like Peru and Mexico, wages have fallen by nearly half, and unemployment has risen sharply. All this would have been worthwhile if the countries concerned had emerged from the crisis and were now back on the road to economic growth. But the size of the debt had become so great that countries had to run faster and faster just in order to pay the interest.

The only exception had been the biggest debtor of all, Brazil, which, partly by some sleight-of-hand with IMF targets, managed to grow by an astonishing 8 per cent in 1986. However, most of these countries' hopes of future growth have suffered; virtually no new foreign lending has come into their coffers to pay for new development projects. The commercial banks have been obliged to lend in order to recycle old loans, but, understandably, have flatly refused to do so for new plant and factories. After half a decade of hardship, the economies of debtor countries have slid back to where they were a decade ago.

Does this matter, except to those who worry about the plight of the poor in the poor world? It should, on a number of counts. First, economic growth in the developing world also benefits the developed world, providing new markets and new opportunities for investment. Second, and much more seriously, the political dangers of allowing the present squeeze to continue, particularly in Latin America, are great. Since much of that continent emerged from under a long spell of far-right dictatorship at the beginning of this decade, Latin American voters have astonished their American neighbours by voting for middle-of-the-road parties; but a drift to the left can now be discerned.

The third threat is to the world financial system itself. Up to now the United States has pursued a 'case-by-case' approach in dealing with the debt problem. That still seems to be working: the Mexicans appear to have won slightly better terms in their recent negotiations with the creditor countries. But as the problem continues, voices from the right as well as the left are calling on Latin Americans to stand together and impose a unilateral limit on the level of debt repayment. This would lead to severe recriminations, maybe including a trade war, on both sides, as well as giving a bad jolt to the world economy. One country, Peru, has already imposed a 10 per cent of exports ceiling on debt repayment and has got away with it to the extent that it has continued to do so for two more years.

Partly to forestall such action, the American Treasury Secretary, Mr James Baker, has been trying to persuade Congress to channel money through world financial institutions like the World Bank and Inter-American Development Bank to assist the debtor countries in growing again. There is an element of this in the Mexican package.

The long-term answer, though, is that the debt burden must be sharply eased, and the sooner the better. The crisis requires a sharing of responsibility between lender and borrower. It could be sorted out if the developed countries were to put their heads together and draw up a plan to allow the debt to be written off over a period of time in such a way as to cause minimal damage to the world economy. This would add a point or two to world inflation, but it need be no more than that.

The potential investors who are holding back from the Third World for fear of default would then feel free to put their money in again. The young democracies of the Third World would have reason to be grateful to the West, instead of harbouring a massive grudge against the international financial system. And, not least, the process of eliminating world poverty through competitive growth could then continue, pulling the rug from under those who argue that the whole crisis is a gigantic capitalist conspiracy and not the mega-economic cock-up it truly is.

Along with the All-Party Parliamentary Group on Overseas Development, I believe there is overriding necessity for new political initiatives:
- a greater involvement of creditor governments in initiating debt relief and adjusting their regulations to enable banks to respond;

- recognition that readiness to write off at least a portion of the principal outstanding would not only give debtor nations a breathing space to restore their domestic savings and investment, but would also reflect the view of the markets, which have already discounted many Third World debts;
- co-ordinated action by the creditor nations, possibly using surplus funds, to reduce the burden of current debt service by returning interest rates to the sort of real levels originally expected by the lenders.

What has been the UK response? Apart from according generous aid-debt relief to the poorest countries, the government have not launched any innovations. Treasury caution has determined the agenda at the expense of our interests in developing countries. The government have preferred a reactive response, closely attuned to the US position. We note, however, that the US itself is now showing a greater sense of urgency. Our banks show realism about the problem, but are awaiting a signal from government. The UK places great trust in the IMF and the World Bank. These are excellent and valuable institutions. But their close involvement in lending and policy prescriptions in less developed countries over the past five years means they are also part of the problem. They must be part of the solution, but the political will to innovate must start elsewhere.

The issue must be treated at the level of governments. All interested parties, debtors as well as creditors, developed and developing countries, must be heard. All the issues, economic and social as well as financial, must be covered. That is why we believe the United Nations to be the most appropriate forum. We propose, therefore, that the UK government now take the debt issue to the United Nations General Assembly. We believe that support should be urgently sought for the spirit of our recommendations from the Commonwealth, our European partners, the United States and Japan. Governments could then take a stronger consensus view to the annual meeting of the World Bank/IMF in September 1988.

10 A Britain Fit to Live In

The environment is the back-burner issue that needs to be elevated to the forefront of British politics, and the Conservative Party is the party best suited to do it. Conservatives are, after all, historically the traditional party of the British countryside; much of their support derives from those who care about the countryside and who care about living conditions in the cities. The Party's objective has been to help all of the British people to become members of a capital-owning and a home-owning democracy. Our objective is also to help them become members of a leisure-enjoying democracy and to continue to improve the climate in which they live and work.

The Conservative Party is better qualified than other parties to do this. The Labour Party's commitment to full employment has traditionally made it put the environment on a minor footing: jobs traditionally come before anything. This also applies to the Social Democratic approach. The Liberal Party, committed to industrial capitalism, showed little concern for the environment up to its demise as a government party in the 1920s; only now, belatedly, has it done an about-turn which takes equally little account of the need for economic growth and job creation.

The Conservative Party is committed to economic growth, and in this is opposed to the eco-parties which urge a halt to it; indeed, we believe that economic growth through technological innovation will provide a much 'clearer', less environmentally damaging, less polluting kind of growth. The Conservative Party roundly rejects, however, the kind of small-minded elitism which suggests that the economy need not provide greater wealth to eliminate the poverty and depression that exists in much of the country. This anti-growth argument is usually advanced by those who can afford to live away from depressed areas. Growth and the creation of wealth are essential; but they can be harnessed environmentally so as to make the fruits of greater prosperity more enjoyable to the majority of people.

For Conservatives the environmental problem is doubly important, since there are signs that the Marxist left, as well as irresponsible ecologists, are seeking to manipulate the issue into one that will extend the state's control into farming, industry and other activities. Conservatives who believe in limiting the role of the stage should fight to ensure that, through the proper and limited measures that are needed to protect the environment, no such excuse is afforded the left.

Britain's special problem is that it is a small, overpopulated country with finite space for building, development and leisure activities. Britain's countryside is a precious commodity, not so much because of its scarcity – in fact enormous tracts of countryside have survived the planning spree of the 1950s and 1960s – but because of the growing demand for it. Rambling as a pastime is estimated to have mushroomed in recent years. The same can be said for many other country pursuits, including recreation on inland waterways, climbing and sailing. These are sharply increasing as ever larger numbers of people have the time and money available to go on self-catering or hotel holidays within Britain. The growth in country pursuits provides two challenges: first, the countryside must be conserved to allow people to pursue these activities; second, the pursuit of those activities must not itself spoil the countryside. Up to now, it is the first challenge that has met the most vigorous response.

The Countryside Act of 1968 did close most of the loopholes that have allowed encroachments into the countryside up to now. Thus, it has become more difficult for urban authorities to take over farmland for building purposes. True, recent legislation allows some inroads to be made on green belt development, but this is largely to clear up anomalies within the green belt area, and the vast majority of the restrictions upon these developments remain in force and must continue to do so. Most local authorities are also stringent in enforcing the rules with regard to new planning applications in the countryside for new buildings, caravan sites, etc.

The development of villages proceeds apace largely in order to prevent the building of eyesores in the countryside; but for the most part this has been discreet, well planned and monitored carefully by conservation groups. Planning consent is required for most improvements and additions, which in the case of the average householder has been a very effective constraint over unsightly development.

The main bone of contention has been between farmers and organisations representing country users. The farming community has been widely portrayed as taking advantage of EEC grants to indulge in a spree of hedge and tree destruction and to erect ugly farm buildings on which no consent is required. These extremes have largely occurred within lowland areas of England, and it is vital that the Countryside Act be toughened to include hedge removal and the extension and erection of buildings beyond a certain size within the scope of the Act. The National Farmers Union policy document of 1984 goes a long way towards accepting farmers' responsibility on this score. A ban should also be imposed on the clearance of wood in areas where this would impinge on the natural character of the landscape, although drainage improvements *per se* are usually desirable.

However, to go on to argue, as many country users do, that there should be an automatic right of access to the countryside and that visitors, because they represent a larger group than farmers, have more right than they to the land is likely to result in more wanton destruction of the land than anything else. For countryside to remain in anything like its present picturesque state, farming must be allowed to proceed with as few inhibitions as possible. That means that visitors must respect the rights of farmers, and not interfere with their activities: the alternative is to turn the countryside into a museum, where farming activities are largely sustained through tourist activity – through the exploitation of visitors. The number of farmers that flout the regulations regarding the siting of caravans on their land is a testimony to the destruction that unplanned tourist activity can cause. Rights of access are already plentiful and well-defined. If visitors respect the farmers' ownership of the land, and behave as though they are on someone else's property, the land will be better conserved.

Coupled with a strengthening of the law regarding planning on farm buildings, there should be a strengthening of the civil action that farmers can take when visitors inflict damage upon their property by, for example, destroying fences or leaving gates open – which often involves the loss of livestock. Too often the farmer cannot apprehend the culprit. But it would enhance farmers' sense of security in the face of the numbers coming on to the land if they had at least the right to seek compensation.

Besides staking out and defining rights of way – as has for the most part already been done – it is hard to ensure that visitors do respect the rules. Part of the answer has been supplied by designation of areas into National Parks, Areas of Outstanding Natural Beauty, and Sites of Special Scientific Interest (SSSI). It is fair to say that the initial mistakes of the conservationist organisations are being gradually sorted out. These bodies tended to introduce too much regimentation on the areas they were designed to protect, putting unsightly signs, parking spaces and litter bins around the countryside. The landscaping of these today is more subtle. The SSSIs have, for the most part, been more of an irritation than was necessary. The designation of the Berwyn Range in North Wales, for example, in order to protect a tiny group of birds, has incensed the local community, and it should be possible to make future designations much narrower and more specific. Some Areas of Outstanding Natural Beauty have again aroused antagonism with little obvious benefit.

The two main problems of the urban environment are ugly redevelopment and dereliction. Britain has some 21,000 architects, more per head of population than any other country, responsible for designing some 85 per cent of all new buildings. Architects, however well trained, have tended to come from poor educational backgrounds, which may explain the lack of imagination and aesthetics in their designs. Architects have been accustomed to providing the cheapest solutions with the cheapest materials available, rather than pleasing buildings. Public bodies have very reasonably tended to go for cheap answers; private developers have done so in order to make fast profits.

Coupled with poor designs was, for a long time, a vandalistic approach to existing architecture. Urban development was rightly frozen in the historic centres of Washington, Rome, Amsterdam, Brussels, Prague and Paris. The result in Rome's case has been urban traffic chaos, but few residents or visitors would argue against the policy. In London's case, and that of other major cities throughout the 1950s and 1960s, breakneck development proceeded with only scant regard to the historic nature of the buildings involved (until historic building legislation was introduced in the 1960s). No attention was paid to skylines or the contiguity of buildings, although planning permission began to be extended to this in the late 1960s. By the mid-1960s, London, for example, had some 1000 buildings over 100 feet high.

The scene has changed, although acts of vandalism – like the erection of a space-launcher utterly at odds with the buildings in St James's Street, or the demolition of Kensington Town Hall, or the erection of a conference centre resembling a chemical plant opposite Westminster Abbey – continue. A disturbingly large quantity of buildings are still scheduled for demolition, but listing of grades one, two and three buildings has helped to stem the tide, and local authority planning departments have grown more responsible about the granting of planning permission for unsightly development.

Local authorities, partly for want of cash, have done nothing to improve the quality of their old housing stock, however, apart from abandoning tower block development, surely the most disastrous idea to be conceived by the planners since the Second World War: most of these were not only dangerous but anti-social, leading to isolation for old people and vandalism of the lifts. The grim concrete spaces built beneath them were deemed too small, which led to an extension of these spaces, invalidating the original purpose of the blocks – to save space. Councils have begun to recognise that the existing housing stock in slum areas needs to be improved, and that these have the advantage of favouring a community spirit which is lacking in the new buildings. The 1979 Conservative government's decision to provide large funds in repairs grants was the single biggest boost to such renovation; this process has also been boosted by the privately funded 'gentrification' of parts of big cities, as middle income groups have moved into rundown housing areas and improved them.

We would advocate one further measure to consolidate these advances. The three listed buildings categories should be merged into two, one of which would apply to all buildings built before 1914, the other category to buildings built before 1945. In order to demolish a listed building, a local authority should have to prove that the building could not, under any circumstances, be put to commercial use in the case of a commercial property, or was unsafe or irreparable in the case of domestic housing. Interiors of special significance should also acquire listed status, although the great majority of interiors could, of course, continue to be altered at the owner's wish.

The problem of dereliction is the more difficult to resolve. The shortage of available funds has made it difficult for government, both local and central, to improve derelict areas: there are some 120,000 acres of derelict land in England and Wales, and some 15,000 in Scotland. Most of the dereliction consists of old slagheaps, old mining workings, old industrial premises, and land that has subsided. Only around 3,000 acres are reclaimed every year, while some 3,500 acres are added every year. Until money is available in sufficient quantities, the problem cannot be dealt with.

One approach might be community schemes, of the kind that we suggest should be manned by school-leavers, with as much funding as resources permit. Nationalised industries are already growing more responsible in this respect, and firms abandoning one site for another (as opposed to closing down) could be made more responsible for the improvement of areas they leave behind. A third approach might be preferential fiscal or regional aid treatment for firms occupying sites of dereliction.

The damage caused by transport pollution is an area that has barely been touched by government, largely for fear of dealing a further blow to Britain's motor industry. The car's impact on the countryside and major cities is now irreversible. Hopes of the introduction of safer, quieter alternatives such as electric cars are still, in mass terms, a distant dream. In 1960 there were 9 million vehicles on the roads; in 1964 there 12.5 million; in 1975 there were 18 million; in 1985 there were 22 million. Road spending rose from £65 million in 1947 to some £325 million by 1968. Today, there are some 1,520 km of road per 1,000 square kilometres.

Many of these statistics, and the complaints about them, are misguided. The actual amount of road only increased from 183,000 miles in 1947 to 200,000 in 1967. Britain has always had a larger network of road than most countries, and although these have been expanded and widened, the idea of roads multiplying and spinning out across the British countryside is wrong. There have been appalling individual examples of roads, and in particular motorways and dual carriageways, blighting the countryside, or being steamrollered through by the authorities, using the expedient of building them in sections, and creating traffic chaos at the end of each section. But for the most part, conservationists have successfully brought the road-building spree to an end.

They have been much less successful in controlling unsightly development that often accompanies road development: hideous petrol stations and garish or grim service stations are the norm rather than the exception. Also there are parts of the country – in particular Wales – where environmentalists have failed to stop local authorities from improving relatively underused country roads and straightening every bend – even though this just leads to motorists driving all the faster.

There is beginning to be agreement among both the planners and environmentalists that the building of certain roads is desirable to relieve congestion and pollution elsewhere. This particularly applies to bypasses, built in order to divert traffic from tearing out the hearts of villages and towns. In many historic city centres, urban motorways and large traffic schemes dominate large areas and create an impression of bleakness that probably cannot be reversed. Again, the main redeeming feature is that the destruction of city centres in order to accommodate new road systems has been stopped.

There are other costs of the car, apart from the straightforward destruction of the countryside, or buildings. One is the cost in human lives and injury – about which, however, most governments have been acutely sensitive. In addition, there is the cost of traffic congestion, which may be as high, in terms of delay and wear and tear, as £1 billion a year. The cost of accidents themselves may be as high as £500 million a year.

The other major effect of the car is air pollution. It is reckoned that nearly six million tons of carbon monoxide, hydrocarbons, aldehydes, nitrogen oxide and sulphur dioxide are emitted by vehicles. More than half of all aerial pollution comes from cars. Britain has been astonishingly slow at adopting regulations to control the level of emission: carbon monoxide, in particular, which can cause nausea and giddiness and, in extreme circumstances, death, is discharged at the rate of 3 lbs per gallon of fuel. A danger level of carbon monoxide inhalation is 50 parts per one million parts of air over an eight-hour day. On city streets, it is reckoned that the level reaches 30 to 40 parts per one million parts of air much of the time.

Yet it is more than apparent that environmentalists calling for a major switch away from the car are crying for the moon. The car is the single most freedom-endowing means of transport for millions of people and provides one of the greatest pleasures, as well as being essential to country living and, for millions of people without access to easy public transport, urban living. The motor industry itself cannot be neglected, employing as it does some two million people (although this figure has dropped sharply from the 2.6 million of the late 1960s).

Much more realistic are hopes of an integrated national transport policy which would serve to ease the burden on the roads and the problems of the inner cities. For this, a proper costing policy is necessary, which takes into account the relative costs of various forms of transport. Competition between road and rail, for example, has had little effect in improving the service or costs of either. This is because railways, being a nationalised industry, tend to decide freight charges in line with the need to subsidise loss-making passenger traffic and with their own curious fare structures, while road hauliers are strictly competitive. The bulk of freight goes by road. Similarly, the cost of road transport still remains so much cheaper than rail transport that it costs almost as much for one person to drive as to go by rail; if he takes a single passenger, the cost is halved. Of course, driving involves wear and tear that train journeys do not. Against that, cars take people from door to door without the inconvenience of struggling with luggage at stations, etc.

We believe that it is time for a more considered approach to inter-city traffic. Specifically, the cost of repairing roads ought to be set against the revenues from road and petrol tax in arriving at an indication of the amount the state subsidises private transport. Tolls, as exist in most of continental Europe, should be introduced on motorways to provide a proportion of their upkeep. This would have the effect of tilting the scales in the unfair competition between road and railway. The environmental advantages of railways ought to be taken into account in composing a new fare structure. Existing experience with cheap fares suggests that it is possible to entice a far larger volume of passenger traffic than at present by sharply reducing fares. For environmental purposes this needs to be encouraged.

The axeing of branch lines is not reversible by the state: a state-owned network seems unlikely to make branch lines pay, although there is plenty of evidence that the axeing of branch lines was itself responsible for a sharp drop in passenger traffic along the main arteries; this had not been foreseen when the economics of branch lines were considered. However, private enterprise, in the shape of railway societies or even small firms, may one day be able to retrieve some of the small branch lines and make them pay as commercial, as opposed to tourist, concerns. It is clear, though, that the railway network is effectively stripped to the bone and that the axeing of even non-paying lines to medium-sized towns would add to traffic congestion, particularly where those towns are ill-served by motorways. Thus we recommend an end to any further branch line closures.

The second major transport problem concerns urban traffic. In the big cities, local authorities have been required to run bus and, in the cities that have them, underground services with minimum subsidisation. In taking account of the level of subsidy, scant regard has been paid to the costs of traffic congestion, wear and tear on the roads, or the pollutant effects of heavy traffic. We believe that, in introducing an integrated urban transport policy, a proper comparison of the subsidies given to cars (through road repairs, etc.) and public transport should be drawn up, to allow a better basis for costing future subsidies to the public transport system. Schemes which allow people single fares for any journey, or even free access to public transport, but which allow savings in manpower, should be drawn up, as another inducement to use public transport. Finally, public services need to become a great deal more reliable and efficient than at present if they are ever to induce transport users away from the roads.

Many environmentalists urge bans on cars in inner-city areas. In Oxford and other cities, limited schemes have been introduced, very successfully, and the pattern should be reproduced elsewhere in towns of the same size. It is much easier, though, to introduce such schemes in compact city centres than in, say, London or Birmingham (whose centre has been largely designed for urban traffic). Another idea is for motorists to pay a tax for coming into the big cities, as a discouragement. The administrative and other difficulties would, we feel, obliterate any possible advantage from such a scheme.

The main discouragement – the expense and difficulty of parking – already exists. One such scheme we did favour was for a relaxation of the licensing laws on taxis, in order to allow more drivers to take on paying passengers: we believe that this would significantly reduce the number of cars travelling to and from work, within, in particular, inner-city areas. A ban on all road-widening in inner-city areas, advocated by some environmentalists, was not felt necessary: a toughening up of the listed building system along the lines already suggested, which would, except in very unusual circumstances, take precedence over development schemes, would be sufficient to protect most inner cities. But there are still many traffic improvement schemes which are of considerable environmental benefit; provided that this is as much a priority for a scheme as smooth traffic flow, it should not be interfered with.

The two other problems are pollution of the air and the water. Air pollution is expensive in its effects: it costs some £300 million a year in corrosion, depreciation on buildings and cleaning costs. The loss of fuel under conditions which produce air pollution is reckoned at around £100 million a year; air pollution also makes workers and machinery less efficient, costing around £110 million. The main causes of air pollution are cars, smoke, grit and dirt; and pollution from power stations. Oil-fired power stations produce carbonic acid, which is bad for paintwork and fabrics; coal-burning power stations produce some 0.3 million tons a year of grit and dust, and some 2 million tons of sulphur dioxide.

The 1956 and 1968 Clean Air Acts were revolutionary in their effect, bringing to an end, except in rare climatic conditions, the appalling smogs so common in London and other major cities before that. The four big smogs since the war killed nearly 7,000 people. The costs of the Clean Air Acts were not daunting, in terms of improved quality coal required for open fires or conversion to other fuels, certainly not in comparison with the benefit; and the legislation has been widely welcomed. We would recommend that all urban areas be made smokeless under the terms of the Clean Air Acts (nine tenths of them already are), and that local authorities have an obligation to prosecute any firms emitting major industrial pollutants into the atmosphere. Such controls add to industrial costs, but they are necessary, and there is no reason why the whole community should pay for a firm's poor industrial production techniques. Higher standards of emission control, like higher safety standards, are a necessary cost to industry. Emissions from cars should also be controlled: stronger standards to control lead and carbon monoxide emission as well as existing tax incentives, should be enforced upon motor manufacturers, similar to those already in force in the United States. If this adds £30 or £40 to the price of a car, then it is a necessary price.

Pollution of water is still a major problem, although great strides have been made towards introducing proper processing plants and in processing industrial pollutants. It has been estimated that, during the late 1960s, there were some 5,000 miles of river that were severely polluted. The river Chelmer in Essex, the River Dee in Clwyd, and the Thames are among those which have been exposed to severe pollution problems over the past two decades. Among the offenders have been distilleries, tar and oil factories, sugar refineries, glucose refineries, gas works, chemical fertiliser plants, detergent factories, petrochemical works, petroleum terminals and paper mills. The problem of hard detergents has been virtually defeated by government action. Farm slurry is a growing, though controllable, problem.

Sea pollution, and pollution of the coastline, are other major difficulties. No oil spill has yet matched the *Torrey Canyon* disaster in 1967; but the possibility of one remains. Small leakages from oil tankers remain routine. Another problem is the dumping of coal waste into the sea, in particular off the coast of Yorkshire near Durham. This has created a long and hideously blackened shoreline. A further problem still is sewage: the sea, with its tides, is the most effective disposer of pollution for any local authority which borders it; but sewage is frequently washed back. One survey estimated that 70,000 coliform bacteria of excremental origin per cubic centimetre of water was a common level in many holiday resorts. Strangely, this does not constitute much of a health hazard; but it does make bathing a less comfortable activity, to say the least. We recommend that new and tough central controls be instituted to compel local authorities to treat sewage before dumping it at sea, and that the penalties for industrial polluters of waterways be increased. If there are complaints about central government trampling upon local authority prerogatives, the point should be made that the controls exist only to make a few culpable authorities introduce the rules which should be the norm for all of them. Local authorities cannot defend their privilege to go on polluting water.

The final area for environmental control is that of consumer protection. Consumer protection should be extended in controlling the use of artificial colours, chemical additives, pesticides, oestrogens and antibiotics. Labelling legislation should be tightened up to make it necessary for food manufacturers to spell out the exact contents of tins on their labels. Such legislation has taken a major step forward lately, and some absurdities are now taking place – such as the banning of scented erasers, even though no child has yet in fact swallowed one (some 60–70 cases of children swallowing marbles have instead been recorded). It should be made obligatory for specific gravities to be attached to beer pumps in public houses. Finally, consumer protection bodies should be set up for all major industries with a much wider degree of participation than existing consumer councils have today in the nationalised industries.

Britain is a long way along the road of controlling the worse excesses of its industrial development. As the leisure society comes into being, however, Britain must develop further in the direction of becoming a land fit for all those with a little time on their hands to live in.

11 A Modern Democracy
Robert Harvey

Britain today is probably less democratic than it was in the 1930s. The British constitution since then has barely developed at all, although society has rushed past it in a way that makes its provisions more remote, unrepresentative and inadequate. This halt in constitutional development, in the deepening of British democracy, marks a complete break with the country's previous tradition of constitutional evolution, which did two things: it kept British government roughly abreast of social change; and it defused the tensions that occur when a political system fails to keep up with social change. Britain escaped the wave of European revolutions of the eighteenth and nineteenth centuries largely because it bowed, through the Reform Acts, to the pressure for political change and the eventual demand for universal suffrage.

These concessions had to be extracted out of nineteenth- and early twentieth-century continental European ruling elites through bloodshed and periods of dictatorship. The United States constitution proved flexible enough, and its leaders wise enough, to permit the same kind of constitutional development that shaped Britain. Most of Western Europe's constitutions date from the immediate post-war period, and are therefore relatively modern and up-to-date. (The French constitution of 1958 is even more recent than that.) The American constitution has evolved: from guaranteeing black minority rights in the 1960s through the Democratic Party's reform of its primary system in the mid-1960s, through the increased power of Congress at the expense of the presidency that resulted from the Nixon years, changes have been made, in a continuing, creative tussle between the legislature, the executive and the judiciary, and between the federal government and the states.

Instead Britain's constitution has remained paralysed since the vote was given to women in the 1920s. Such constitutional innovations as there are have been confined to a reform of local government in 1972, which arguably exacerbated the existing problem; and the creation of parliamentary select committees, which have been a very welcome, but very inadequate, step in the direction of increasing legislative scrutiny of the executive. Meanwhile, society and the role of government have developed dramatically. The same 600-plus MPs that watched over a government machine that employed fewer than 1 million people in 1930 and was responsible for 18 per cent of GDP in 1930 now watch over a government machine that employs some 6.5 million people and is responsible for 44 per cent of GDP. The role of the constituency in most MPs' lives has increased dramatically. Yet their facilities, powers and practices have not changed to meet these challenges. The result, by and large, is an executive – consisting principally of one of the best civil services in the world, for all its faults – that is unregulated by the representatives of the people. Some people think this is a good thing: that Whitehall experts are better qualified and experienced to order the complex affairs of the modern state than the politicians. To some extent this is true, but Conservatives believe that it is the politician's role to ensure that the needs of the modern state are tailored to what society wants. That is what democracy is about.

The say of ordinary people in modern British democracy is limited, and limited to three functions. First and foremost – and it is the best feature of modern British democracy – there is the power to 'throw the rascals out'. The two-party system allows British voters the power to dismiss a government of which they disapprove. They may not like the alternative much, either; but they can repudiate the incumbents. That is real electoral power.

The second function – in practice a very minor one – is that voters can affiliate to one of the political parties and, through achieving a say in the selection of parliamentary candidates, can make their voices felt. This is a limited power, because so few voters bother to do so that constituency parties are, in practice, controlled by a minority of activists. The third function that ordinary voters exercise is still more indirect: their elected representatives can, when a party leader retires or is pushed, choose his or her successor (in the case of the Conservative Party the elected representatives of the people, the MPs, have an absolute say, although they have none in the leader's choice of cabinet or shadow cabinet; in the Labour Party some rather dubious bodies – the constituency parties and the trade unions – also have a large say, but MPs elect the shadow cabinet).

Beyond that the British voter and his elected representative exercise astonishingly little power. The Prime Minister of the day chooses his (or her) ministers; the ministers head the government machine. Strategic objectives and, occasionally, the direction of a particular ministry under a particularly strong-minded minister, can be changed by the government of the day. The government machine, however, exercises enormous continuity and administrative power beneath that; and Parliament, supposedly sovereign, usually rubber-stamps the decisions of the executive.

Again, there are those who argue that this is no bad thing: any political system boils down to an institutionalised clash between the demands of governability and representativity. Those who run a country must be able to do so efficiently, decisively and swiftly. Yet Conservatives, and indeed any democratic party, believe that the representative organ, the legislature, should be able to ensure (a) that the country is run in society's, not the bureaucracy's, interests; and (b) that the excesses and mistakes that inevitably occur unless government is adequately monitored, should not be allowed to happen. Certainly Parliament's power is too little today.

Indeed, there are four ways in which Parliament's power has shrunk. First, owing to the extension of the franchise and the advent of mass politics, the electorate has come to view the political scene through the prism of the mass media. This has strengthened the power of the central parties at the expense of individual members, because it is the party, rather than a local figure, which voters now put their crosses by.

This perception, more than anything else, gives the whipping system its power. Members of Parliament are loath to rebel, not because of the rather primitive coercive measures used by the whips to try and ensure subservience, but because they are conscious that they were elected as members of the Conservative and Labour Parties, not as Fred Bloggs or Margaret Snooks, and that they are betraying their electors if they go against their party too often; older members with personal followings are the most prone to rebellion. Party discipline, although slightly eased in recent years, is still very tight compared with what it was in the nineteenth century or in the first three decades of the twentieth, before the advent of mass politics. The power of the executive has been immensely strengthened.

Another significant reason for that power is the 'payroll vote', of around 100 members who have government jobs of one kind or another; add to them the 100 who want jobs, and the pool of potential rebels is small. The way in which the executive is drawn from the ranks of the legislature makes Parliament vastly more manipulable by the executive than the United States Congress. (The House of Representatives is elected every two years, and a third of the Senate is chosen every two years; this means that the number of those being dragged in on a President's coat-tails, and therefore loyal to him, is small. In Britain all members of the governing party are to some extent dragged in on the Prime Minister's coat-tails.)

Another reason for the decline in the power of Parliament is the antique nature of the House of Lords. A number of reasons are advanced against its reform, including the fact that, as it is non-elected, it is more independent, and that it has more specialist expertise than the House of Commons. This is probably true of any collection of heterogeneous individuals randomly picked (which is what the hereditary system amounts to). The nominated members are not to be despised: most have performed good service and have considerable expertise; but their age usually makes them ineffective champions of parliamentary power.

The presence of the hereditary element, in a country which no longer accepts heredity as a credential for government, makes the second chamber virtually redundant. It can delay legislation and get the Commons to reconsider the dottier pieces of legislation. It cannot, ever, obstruct a Bill the government wants to get through. Britain in effect has a unicameral Parliament – which is almost unique in the developed Western world.

There are two arguments for changing this state of affairs: first, the executive deserves to be checked more than it is, and by a more independent body than the House of Commons has become; second, that if reform of the House of Lords is not carried out by a Conservative government, it may one day be the only check to the constitutional excesses of a Labour government, and be swept away as a constitutional anachronism by such a government, installing *de jure* one-chamber rule.

A third reason for the decline in the power of Parliament has already been alluded to: the spread of central government and its functions. It is quite impossible for Parliament to monitor any but the widest legislative decisions and any but the most important executive decisions. The increase in the powers of select committees, inspired by the American model, is welcome, but still wholly inadequate. Select committees have tiny research staffs, usually consisting of a clerk and his assistants. They have few powers to extract secret information from the Whitehall machine, and treat civil servants and ministers with a civility and deference that is astonishing to congressional observers. They very rarely hold interrogation sessions in secret. This rather flimsy check on the executive at the top is matched lower down by an absence of democratic controls and by the failure of local councils to scrutinise adequately the workings of local authority executives.

Finally, the doctrine of ministerial responsibility, by which the minister is theoretically responsible for the misdeeds of his civil servants, has degenerated into a laughing stock. The ministers cannot realistically be held responsible, because only one in a hundred decisions taken by a department reaches the minister's hands for serious consideration in a modern state. But the fact that the civil servants cannot formally be held responsible means that no one is, which is thoroughly unsatisfactory all round.

The fourth reason for the decline in the power of Parliament has been its stubbornness in clinging to time-honoured practices despite the changes all around it. Most Members of Parliament lack the time to give more than cursory attention to the issues before them, because the demands of their constituents have vastly increased. Very few MPs, for example, do not have surgeries; very few are not deluged by constituency mail, averaging around 150 letters a week. Very few do not have much more demanding constituency parties than before, which expect them to be in their seats most weekends.

The ancient practices of the House provide, very properly, for filibustering by the Opposition (often agreement is reached through the 'usual channels' between the whips of the major parties as to how long a filibuster will last and how vociferous the opposition to a particular Bill will be); but it is the norm, rather than the exception, that debates, and votes, break not just the 10 o'clock barrier in the evening but continue until at least 11.30; and on one or two nights a week, debates can stretch well past, towards two or three in the morning. The result is that many MPs' energies are expended on a rather ritualistic part of their activities – sitting around and waiting for votes, rather than research, speech-writing or investigating the abuses of the executive. In addition, the relatively low pay of MPs means that many take second jobs, which further reduces the attention they can give to the issues, and means that many people with outstandingly saleable skills in the employment market are unwilling to seek election, resulting in a lower calibre of MP. The scrutiny of legislation is consequently pretty inadequate.

Some legislation has a strongly political content: its general lines will then usually be drafted in cabinet committee. Most laws are drafted by the civil servants with the minister responsible taking only a broad interest in its general lines. A Bill will usually be only cursorily discussed in cabinet. It will then reach the floor of the House and be passed on a whipped vote. In committee, ministers will generally defend the legislation down to its last jot and tittle, although most of it is not drafted by them – indeed, may be incomprehensible to them.

A handful of amendments usually go through, just to persuade the Opposition to drop a particular filibuster. The ministers defending are given briefs by civil servants. Largely unchanged, Bills will then go back to the floor of the House where, at report stage, fresh amendments are usually defeated on a whipped vote. The Bill then goes to the House of Lords, where it is usually amended a bit more, because the whipping is less tough. The amendments may then be rejected or accepted on the floor of the House, again on a whipped vote, before the Bill gets its third reading. The whole process takes up a great deal of Parliamentary time. But very little usually happens to a Bill: the overwhelming majority of its provisions will have been drawn up by, and reflect the views of, non-responsible civil servants. For a Bill that is largely a departmental measure, rather than a politically inspired one, virtually the whole measure will have been the creature of civil servants.

The legislature has very little power in Britain, too little, in our view, to justify the executive's contention that its preponderance is necessary for the smooth working of government. On the contrary, government is frequently tripped up by departmental mistakes, because Parliament has no power to check silly decisions; by poor legislation, because Parliament has so little power to amend it; by irritated select committees, who, denied the information to make evidence-based criticism of the executive, do so anyway as an expression of independence and impotence.

This system is highly convenient for the civil servants, but it is not largely their doing. The responsibility lies, rather, with MPs, who regard any attempt to improve or modify these practices as an infringement of their rights. A majority of MPs probably think the system ridiculously outdated; but the minority has all the normal battery of parliamentary powers of obstruction at its disposal.

The inefficiency of the legislature at the centre is multiplied in the case of local government. Local elections are bedevilled by a low turnout: on average, this is some 35 per cent, compared with a turnout of 78 per cent for parliamentary elections. The shortage of people willing to serve on local government, which is unpaid (although expenses are available and sometimes generous), means that applicants are often of low calibre or have a vested interest in serving in local government. Most often councillors are fairly ill-qualified amateurs.

The reorganisation of local government in 1972 resulted in the setting up of three effective tiers of local authority: community councils, which count for nothing; borough councils, which represent the traditional towns and whose powers are limited (the low calibre of those who serve on them, and of the executive officers – in the latter's case because of the limited pay and responsibilities of the job – is notorious; and the county councils. The county councils are the Rolls Royces of local government, with considerable spending powers and spanking modern headquarters. The size of county council 'precepts' (budgets), the number of their staff and the pay offered attract high-calibre, expert senior staff who, generally speaking, run rings around local councillors. In few places does the finance committee have the expertise to shoot down the spending proposals of the finance director and his staff.

The big metropolitan authorities which were set up to run some of the big cities represent worse versions of the same problem. The councils' impulse has been to increase their spending in recent years from £33 billion in 1981–2 to £38 billion in 1987–8. Central government has sought to curb spending, both because it is undergoing retrenchment itself and because it disapproves of the way councillors elected by a third of the vote and very largely steered by the executive staff, are getting away with overspending. Central government has tried to reduce rate support grant to overspending authorities; and, when this failed, to introduce selective rate limitation powers to prevent councils simply passing the burden on to the ratepayer.

The councils have been inhibited in their resistance to this by the fact that they represent so few people. The apathy that council elections induce is a gash at the heart of British democracy, and one that urgently needs correcting. It is feeble to say that low turnouts are evidence that most people are satisfied, or at any rate not dissatisfied, with the system. Few ratepayers are satisfied, but faced with a daunting choice of unknown names or a straightforward political division, are not prepared to make their views felt. No democrat can be satisfied with British local government as it works today.

The answers to these problems suggest themselves. The first one is the thorny issue of reform of party selection procedures. Both major parties jealously guard the independence of their procedure and insist this should lie outside government control. Yet, on the argument that trade unions are too important to the country and their membership not to be interfered with by the government, the present government has passed a law regulating their election procedures. There should at least be certain minimum provisions governing the procedure of political parties: for example, that all constituency party members be allowed to elect the selection committees and that (as already happens in the Conservative Party) candidates be submitted for ratification by adoption meetings of the whole constituency association – at which challengers may be eligible to stand. The American-style primary system would probably take the activism of constituency party members too far. The main purpose of reform of the selection procedures ought to be to prevent the infiltration of selection committees by small activist minorities, as is happening in the Labour Party today.

Much more far-reaching change is necessary to strengthen the checking power of Parliament in its dealings with the executive. Firstly, scrutiny of the executive needs to be increased. This can be achieved by a reform of parliamentary procedures to sweep away the many purely ceremonial bits of legislation that come before it – private bills, orders in council, etc.; by providing a proper research back-up to the standing committees that examine Bills going through Parliament; by providing proper research and powers of interrogation for select committees examining ministers and civil servants; and by lengthening debates on issues of pressing national importance.

Parliamentary time should be reformed so as to allow sittings in the morning and a cut-off time for debate in the evening, thus easing the unnatural pressure on MPs that contributes to the overhasty scrutiny of legislation. MPs should also be paid about twice what they get now, have their secretarial arrangements managed by a cost-effective central staff, be forbidden from taking second employment, get modern separate offices, and have pay increases in line with inflation.

Proportional representation is often suggested as a panacea to the adversarial yes-you-did, no-I-didn't system of parliamentary debate that is such an obvious defect of the present system. Proportional representation would, it is suggested, allow the political centre to be represented to the full extent of its vote in the country and prevent the country lurching to extremes. In practice, these benefits are probably overstated: the fact that centre party support tends to be smaller in a marginal seat suggests that, given a choice between two parties with a chance of winning, voters prefer to opt for one or the other. The Liberal vote is much higher in safe Conservative seats, where it forms a safe protest vote, and the Social Democratic vote higher in Labour seats, where the same is true.

In democratic terms, it is possible to argue that the introduction of proportional representation could result in a multi-party system where governments are made and unmade through backroom deals, as opposed to the British system where they are installed and unfrocked by democratic vote. Moreover, any kind of proportional list system would dilute the contact between the MP and his constituent that is such a long-standing feature of Britain's constitutional system.

All the same, any system which allows (a) a party polling more than a quarter of the votes to win a derisory handful of seats and (b) a party which does not even have a relative majority of the votes at an election to win power (as happened when Labour formed a government in February 1974) certainly needs reforming. On balance the authors lean towards a system of run-off elections like the French one. This retains the advantages of the constituency member system, but compels voters to make a conscious choice between likely winners (most proportional systems require preferences, which may yield results voters had not originally intended). It also usually results in clear-cut electoral decisions, not backroom dealing. The usual argument against the run-off system is that Britain's voters, only three-quarters of whom vote at general elections, will not drag themselves to the polls a second time two weeks later. The French experience suggests that run-off elections between two candidates actually heighten voter interest, and result in turnouts at least as high.

However, as already argued, it is not the electoral system that is at fault so much as the failure of the legislature to check the executive. Reform of parliamentary procedure is only one element of this: another element should be the power of consumer councils to monitor the workings of the nationalised industries.

Much the biggest element, however, would be the reform of the House of Lords. If a second chamber were set up with a different constituency but with nearly as much legitimacy as that of the House of Commons, and of the government, then the legislature would obtain real power and independence. There are three obvious possibilities, given that a hereditary chamber lacks any such legitimacy: (a) a nominated body; (b) a body elected at a different time to the House of Commons – say a third of its members every two years, as the American senate is, under a proportional system; or (c) an indirectly elected body chosen, for example, to represent the regions of the United Kingdom.

The disadvantages of a nominated body are obvious: nominations would be made by the executive, which would weaken the power of Parliament to check it. Nominated bodies also carry very little legitimacy in the eye of the public, and in a confrontation between an elected chamber and a nominated one it is quite certain who would win. A differentially elected chamber would have the advantage of greater legitimacy, because it is elected; of changing gradually, if its members were elected in staggered elections, thus checking the sharp lurches of fortune that take place in the House of Commons; and, if elected on the basis of proportional representation (along with the suggested run-off system of election to the House of Commons), of defusing the pressure for proportional representation in the House of Commons. An indirectly elected body would, again, lack the popular legitimacy to stand up to an executive rooted in the elected House of Commons. This points towards the second solution.

However, continuity is vital to the British tradition of government, and to Conservative thinking especially. Existing peers should continue to carry titles and be eligible not to sit in the upper house, but to choose representatives to occupy 10 per cent of the seats there. There could also be a nominated element of 10 per cent, reflecting the major interest groups that already sit in the Lords – former members of the House of Commons, judges, bishops, distinguished figures from industry and the trade unions, and even the arts and entertainments world. The remaining 80 per cent would be elected, which would endow the chamber with nearly, but not quite, as much legitimacy as the House of Commons. The best system of election would be by region, on a regional proportional representation basis, with an equal number of representatives per region, to give them the sense of identity they have been crying out for.

It is vital, however, that the new upper house remains subordinate to the House of Commons, because that is the essence of the British parliamentary system. In the United States the Senate has acquired a blocking power, which, particularly in the field of foreign policy, renders decisive executive action difficult. The House of Commons should remain the sole initiator of legislation; the upper house should have checking and supervisory policy only. The upper house should have the power to debate, but not to override the House of Commons on foreign policy matters. On issues of domestic policy, where a clash takes place between the two houses, it should require a three-fifths majority of the upper house to override a simple majority of the lower (two-thirds is too stiff). The government of the day should retain the right to dissolve the lower house if annoyed by the rejection of a Bill, and resubmit a law after such a dissolution, after which the law would not require second chamber approval. In practice this would take place rarely, if ever. The upper house would thus have a strong checking power, but not an absolute blocking power.

It has been suggested that referenda might be used to resolve a deadlock between the two houses. In the view of the authors, referenda have little place in Britain's representative parliamentary tradition: they might well suffer from apathy; they are expensive; they are open to interest group manipulation – groups backing them could campaign and muster their supporters on a low turnout; and they are mere snapshots of public opinion at a particular time. A two-chamber parliament of the kind suggested would reinvigorate Britain's tradition of representative democracy – of leaving the business of governing to the politicians for five years, with the option of throwing the rascals out if they do too badly. Nevertheless, Parliament being sovereign, the House of Commons would retain the right to hold referenda on special occasions if it so wished – as happened with the referendum on entry into Europe.

The regional system of election to the upper house provides an indication of how to deal with the vexed problem of local government. Most people, appalled by what happened in the last shake-up of local government, are opposed to further change. Yet this leaves a manifestly unsatisfactory situation. Local government today falls between two stools: local councils are too small and unrepresentative to provide a check on central government, but are too large and impersonal to arouse much interest among the general public. The borough and community councils are closer to the ground, but in the boroughs' case have too few powers, and in the community councils' case have none at all.

Britain does, however, naturally divide into regions, and people do feel part of a region. Regional assemblies would bring government closer to people – which is especially important as very remote EEC institutions gain power. As paid mini-parliaments, with powers to levy local income tax and to spend on a wide range of social services, regional assemblies should attract as much comment and electoral participation as the Westminster Parliament. They would also perform the invaluable function of checking the actions of central government officials at a lower level than the Westminster Parliament can possibly do.

Britain divides into obvious regions: the South West; the South Centre; London; the South East; East Anglia; Central England; the West Midlands; South Wales; North Wales; Lancashire; the North Centre (Nottinghamshire, Derbyshire and Lincolnshire); Yorkshire and Durham; the North West; Clydeside; Edinburgh and South Scotland; North East Scotland; North West Scotland. Only one other tier of local government is necessary: that which is most closely modelled on accepted existing communities. In the countryside this would be villages and small market towns; small cities would retain borough councils; in the big cities the second tier would be modelled on existing boroughs. Their powers would be largely the same as existing borough councils, and those who sat on them would not be paid.

It is thus proposed to do away with one tier of local authority, the metropolitan councils, and to merge another two, the county councils and the borough councils. Participation in elections for the lower tier would probably be much lower than that for the regional assemblies; but they would serve the need for councils closer to people. Such a system would go a long way towards defusing the pressure, now gaining strength, for devolution in Scotland (and the much lesser pressure in Wales).

A written constitution is not required to set out the powers of local government versus central government, or of the two Westminster chambers. Each will gain authority through its democratic legitimacy. An unelected House of Lords can be railroaded; a barely elected local council can be railroaded; but elected assemblies cannot be ridden over roughshod by the executive acting through the House of Commons, in the way that happens now. A written constitution would freeze constitutional arrangements, while providing a field day for lawyers and paralysing the processes of government, as to some extent the American constitution does. The sovereignty of the House of Commons in Parliament must remain paramount, because it provides a final source of decision-making; a judiciary should not assume that role.

Change will only come about over a protracted period. Yet we have set out, we believe, a blueprint for constitutional reform based primarily on the establishment of an effective second chamber and of effective regional government. We are concerned that without progress towards these goals, Britain's constitution could well become more entrenched still, ever more bypassed in a changing society, and in failing adequately to reflect the needs of that society, will one day be roughly, even violently, discarded. Conservatives believe that gradual change is vital to defuse pressure for sudden change.